T0229868

GLACIAL INDICATOR TRACING

# GLACIAL INDICATOR TRACING

*Edited by*
**RAIMO KUJANSUU & MATTI SAARNISTO**
*Geological Survey of Finland, Espoo*

A.A.BALKEMA / ROTTERDAM / BROOKFIELD / 1990

Published by
A.A. Balkema, P.O. Box 1675, 3000 BR Rotterdam, Netherlands
A.A. Balkema Publishers, Old Post Road, Brookfield, VT 05036, USA

ISBN 90 6191 857 X

# Contents

# Preface

The present volume describes the methods employed in glacial indicator tracing, which was the aim of Work Group 9 of the INQUA Commission on the Genesis and Lithology of Glacial Sediments entitled 'Glacigenic deposits as indicators of glacial movements and their use for indicator tracing in the search for ore deposits'. The Work Group emphasised the significance of extensive application of glacial geology, including both the lithology and geochemistry of glacial sediments and their genesis, stratigraphy and direction and distances of transport. The aim of the present volume is to serve as a final report of the Work Group and as a handbook of glacial indicator tracing.

We are grateful to Aleksis Dreimanis, President of the INQUA Commission on the Genesis and Lithology of Glacial Deposits for his constant support during the planning and editorial stages of this volume; to Michel Bouchard and Veli-Pekka Salonen for reviewing some of the articles; to Malcolm Hicks for his translations and linquistic checking of a number of the manuscripts; to Maila Koivisto for the typing; to Satu Moberg and Hilkka Saastamoinen for drafting many of the illustrations, and to Petri Lintinen, M.Sc. and Saakko Putkonen, M.Sc. for technical assistance in editorial work.

Espoo, 1st February 1989

Raimo Kujansuu
Matti Saarnisto

# CHAPTER 1

# An outline of glacial indicator tracing

MATTI SAARNISTO

*Geological Survey of Finland, Espoo*

## INTRODUCTION

Glacial indicators are particles of any size transported by glaciers (Milthers 1909). They form indicator trains or indicator fans trailing down-glacier from their source rock, or provenance. These were earlier called 'boulder trains', a 'boulder' being a clast of any size (for the history of these terms, see Flint 1971). An erratic is a term applied to a particle of any size contained in drift or lying free on the surface. It is used mainly for particles transported by glaciers – as in the present volume – but it is not confined to these alone. Glacial dispersal is the process of transport and deposition of glacially eroded material (e.g. Shilts 1984). The term boulder tracing is used in mineral exploration for clasts of any size, although mostly for true boulders, of at least 10 cm in diameter (see Bouchard and Salonen, this volume).

The present volume describes the methods employed in glacial indicator tracing, including not only boulders but also the general lithology of glacial sediments, their heavy minerals and geochemistry. The methods available to glacial geologist for ore prospecting have been used successfully in areas such as Finland, where the glacial sediment cover is thin or of moderate thickness (e.g. Kujansuu 1987, Saltikoff 1984). However, as the easy targets have gradually been discovered, exploration has became more and more demanding and should be based on sound scientific work, including understanding of the ore potential of the bedrock, its geophysical properties, characteristics of the glacial sediment cover and postglacial hydromorphic alterations in its geochemistry. Currently the search for ore is often based on weak, poorly recognizable or isolated indicators, or targets which seldom produce traceable boulder trains or unambiguous geochemical anomalies. In these cases detailed glacial geological investigations are needed. The genesis of glacial deposits, their lithofacies and transport characteristics, have to be understood, as underlined by Bouchard and Salonen.

Problems of indicator tracing are different in the marginal areas of former ice sheets, where the glacial sediment cover is thick and the proportion of long-distance material can be considerable (Stewart and Boster, this volume). Similarly alpine environments serve as a special case in indicator tracing due to the large amounts of glaciofluvially transported material and often clearly defined provenance area (e.g. Stephens et al. 1983, Evenson and Clinch 1987), and problems in permafrost areas differ from those in unfrozen terrain (Shilts 1973 b, see below).

The aim of this outline of glacial indicator tracing is to review and summarize the topics of the book. Special emphasis will be given to investigations dealing with till, the most wide-spread and commonly used material in glacial indicator tracing. Information from an ore outcrop is widely distributed in till or other glacial sediments, far exceeding the area of the outcrop itself, although it is also diluted and often difficult to recognize. The indicators are widely dispersed in glacial sediments, which offer a unique media for successful exploration work and should not be recarded by any means as a hindrance.

A single ore boulder is an indisputable sign of a mineralization somewhere, but anomalous values in soil geochemistry cannot be interpreted in such a straightforward way. Many aspects of geochemical drift prospecting have been touched upon by Shilts (e.g. 1973 a, b, 1975, 1982, 1984, Shilts and Kettles this volume) and Coker and DiLabio (e.g. 1989) which can be directly applied to the whole field of glacial indicator tracing. The following discussion greatly benefits from the above articles.

GLACIAL MORPHOLOGY

Glacial landforms are important for reconstructing directions of glacial movement. They can be investigated using air photography (Kujansuu, this volume) or topographic maps provided that these have adequate contour intervals. In his article on glacial morphology as an indicator of the direction of glacial transport, Lundqvist nevertheles emphasizes that landforms, although of great importance, should not be used alone, but should always be combined with an investigation of the stratigraphy, lithology and till fabric and complementary studies of till genesis and glaciodynamics. This points to the need to use all glacial elements in indicator tracing.

The landforms considered are of varying magnitude, from major glacially sculptured bedrock forms to glacial striae. In between these lie the landforms of glacial deposits, both morainic and glaciofluvial. The former include radial forms such as drumlins and flutings and transversal forms such as Rogen (or ribbed) moraines and De Geer moraines. Glaciofluvial accumulation forms are not such accurate indicators of glacial flow as are striae and radial streamlined moraine forms, but the directions of eskers in particular can give information on the regional ice flow pattern and reflect the very last ice movement very sensitively in places, a movement also indicated by striae and the till fabric. In an example from the Riihimäki area in southern Finland, Kujansuu shows how precisely the directions of eskers and De Geer moraines reflect late changes in the ice flow.

The relationship between glacial dynamics as indicated by moraine morphology and boulder transport is discussed by Bouchard and Salonen (see also Salonen 1986). The average transport distance for boulders was between 1 to 5 km in ground moraine areas and between 7 and 20 km in drumlin areas. The material in the boulder-covered surfaces of active-ice hummocky moraines has travelled only between 0.5 and 2.0 km on average.

According to Stewart and Boster (this volume), glacial landscapes at the marginal areas of continental glaciers are characterized by end moraines and related landforms. They consist of tills and outwash sediments whose genetic diversity precludes the use of moraines for more than a general assesment of the regional provenance of a glacial lobe. There are some general trends to be detected in the provenance of the tills, however. The basal zone of a till unit will be especially enriched with local material, and older tills deposited close to bedrock will similarly contain greater amounts of underlying rocks than do the successively younger tills stacked on top. The till matrix (less than 0.063 mm fraction) in the marginal areas of the Laurentide ice sheet, and probably also the Fennoscandian ice sheet, will consist predominantly of two general lithological components: (1) carbonate minerals and phyllosilicates derived from (local) Phanerozoic rocks, and (2) silty quartz-feldspathic rock flour eroded mainly from Precambrian rocks. Thus exploration in marginal areas is hampered by the often long-distance transport of the material in the till and the great thickness of the glacial sediments.

TILL AS A MATERIAL FOR GLACIAL INDICATOR TRACING

Till is the material most commonly used for glacial indicator tracing. A theoretical background to the dispersal of glacial debris is given by Puranen in his article entitled 'Modelling of glacial transport of tills' (c.f. also Puranen 1988). The model is applicable to areas of thin till cover, as also are most of the methods used in glacial indicator tracing. Till genesis is important as far as glacial transport distances are concerned. Basal debris will have travelled only short distances as compared with englacial or supraglacial debris, and the bottom parts of till beds are more local than the upper parts. Especially long transport distances are found in the case of surficial boulders, i.e. erratics, with examples from North America and Europe involving distances of more than 1500 km (e.g. Flint 1971, Prest and Nielsen 1987). Surficial boulders can also be transported by icebergs in proglacial lakes or seas. The provenance of such erratics is difficult to assess, and therefore they form a special problem in mineral exploration even in areas of a thin till cover, and especially where single, isolated boulders are concerned.

The formation and deposition of subglacial and supraglacial tills, familiar to most prospectors as basal and ablation tills, is dealt with by Dreimanis. From the prospector's point of view, those till properties which contain information on predepositional transport (such as fabric) or on the derivation of ore fragments from their bedrock source (such as lithology) are of particular importance. It is easier to decipher such properties if the position of deposition of the till in relation to glacier ice is correctly identified, and also if the most probable thermal and dynamic regimes prevailing in the till-depositing glacier are understood. Dreimanis summarizes in detail the descriptive properties of different genetic varieties of till which are useful in practical field work.

*Till texture and lithology*

Till is composed of rock fragments and individual minerals. Rock fragments predominate in coarse fractions but in finer fractions of less than 0.5–1.0 mm the material is mostly monomineralic. Every rock type in till tends to have a bimodal distribution, being coarser in the clast fraction and finer in the till matrix. The clast mode is larger when the material is of short-distance origin but the matrix mode increases as a function of increasing transport distance due to comminution in transport (Dreimanis and Vagners 1971 a, 1971 b).

The original grain size of the till forming rocks also influences the texture of the till matrix. Fine-grained rocks such as limestones and siltstones produce a silt and clay rich fine matrix, and therefore the tills in areas of such rocks are fine-grained, whereas the coarse grained metamorphic and igneous rocks remain in the coarser fraction resulting in a silty and sandy matrix (e.g. Jørgensen 1977, Haldorsen 1983).

Differential resistance of rock types to glacial abrasion influences the transport distance of the various grain-size classes of the till. The tendency for different minerals to become enriched in certain grain-size fractions during glacial transport has to be considered when tills are studied from the prospecting point of view. Tough, tenacious rock types are transported longer distances in the clast and boulder fractions, while soft ore minerals such as sulphides are comminuted to the finest fraction (finer than their original grain size) within a short distance (e.g. Nevalainen 1989). Erosion resistant hard minerals such as oxides are enriched at their original grain size and in this fraction transported long distances. When analysing trace metals, an adequate fraction should be chosen separately in each case in order to obtain the best results, as will be discussed below.

To sum up, when considering the transportation of an ore boulder or of fine till fractions containing material from ore minerals, the properties of the parent rock should also be considered in addition to glaciological and stratigraphical questions.

*Transport distance v. grain size, a discussion*

It has been concluded in many studies on the transport distance of various till fractions that the fines, or at least the silicate material contained in them, represent glacial debris from much wider areas than the coarse fractions. This has been demonstrated in Finland by Virkkala (1971), Perttunen (1977), and Saarnisto and Taipale (1985) and Taipale et al. (1986), in Sweden by Lindén (1975) and in North America by Harrison (1960) and Dreimanis and Vagners (1971 a, 1971 b). Further references dealing with the lithological homogeneity of tills include Shilts (1971, 1973 a), Gross and Moran (1971), Gillberg (1965) and Svantesson (1976).

There are also some studies, however, especially based on till geochemistry, in which the fine fraction has been interpreted as having been transported over shorter distances than the coarser fractions. In Finland for example, Kauranne et al. (1977) and Salminen (1980) considered the fine fraction more local than the coarse fractions when studying the transportation of easily weathering sulphide minerals which are comminuted rapidly in the course of glacial transport. This apparent discrepancy will be discussed below.

Sorting of glacial debris by grain size as a consequence of englacial transport so that finer fraction travels shorter distances than the coarser material is difficult to envisage, whereas the apparent longer transport distance for till fines can be explained by further comminution of coarse debris during glacial transport once the feeding of debris of a given lithology has ceased. Boulton (1975) observes that particles transported in suspension, that is entirely within the ice, move at the same velocity as the ice itself, but some sorting by grain size may occur at the base of the glacier, in the zone of traction, because grain size influences the velocity of particles that are in contact with the bedrock. The intermediate grain sizes, i.e. coarse silt, sand and fine gravel, move faster locally than do the fine or coarse fractions.

GLACIOFLUVIAL MATERIAL

The transport distance is generally longer in glaciofluvial material than in the till from which it is derived (from kilometres to tens of kilometres, see Lilliesköld, this volume) and thus the source area is more difficult to assess. Sorting by water and clast weight further complicates the picture. In poorly known areas, however, gravel pits in eskers offer a rapid, easy way of gaining general information on the bedrock on a regional scale.

Eskers often follow zones of structural discontinuity in the bedrock, such as faults, an observation that may indicate that eskers are preferentially located over sites with a high potential for mineralization, as pointed out by Shilts (1984). Thus eskers should not be overlooked in indicator tracing although the great random variations and anomalous trace metal contents of hydromorphic origin (Fe-Mg precipitates) indicate that glaciofluvial material is not easily comparable with till in this respect. In addition, absence of the fine fraction and leached sulphides and carbonates contribute to generally low trace metal concentrations. Eskers are widely spaced and their material represents a zone of unknown width along their length, and they are therefore suitable only for reconnaissance studies and seldom for detailed sampling (c.f. Coker and DiLabio 1989).

There are cases, howewer, in which esker material has been useful in mineral exploration, as shown by the discovery of a kimberlite belt in Canada (Lee 1965, Lilliesköld, this volume). Other references include Baker (1982) and Martin and Eng (1985). The usefulness of esker material for mineral exploration in permanently frozen terrain in the Keewatin area of Canada has been demonstrated by Shilts (1973 b). There the coarser esker material has undergone less secondary chemical alteration than the mud boils which have developed in the till due to frost action.

HEAVY MINERAL INVESTIGATIONS

Heavy minerals are very useful for tracing till provenance, examining till stratigraphy and prospecting for ores, and their use in systematic drift prospecting has increased in recent years. Peuraniemi has summarized the field and laboratory procedures used and presented a number of case histories. Heavy mineral investigations succeed best when the minerals used are resistant ones such as cassiterite, chromite or scheelite, although more easily weathered minerals such as sulphides can also give good results. Where boulder hunting does not necessarily give an indication of the presence of an ore, this can be obtained by studying the heavy minerals.

The transport distances for heavy minerals, i.e. distance from the source to the point where the minerals can be found in anomalous amounts, may be several kilometres. In the eight Finnish cases reported by Peuraniemi the average was 6 kilometres and the extremes 0.5 and 17 km. This is usually very much longer than for detectable geochemical anomalies in the fine till fraction. The heavy mineral anomalies can usually be found using a less dense sampling grid than for trace element anomalies in till fines, so that the method can be used to survey vast areas quite quickly. The analysis can be performed to some degree in the field by conventional or mechanical panning and thus the results can be used to guide ongoing exploration work, as pointed out by Hirvas and Nenonen. A heavy mineral inventory does not replace boulder tracing and geochemical investigations but can rather be used to complement these.

DRIFT GEOCHEMISTRY

Drift geochemistry is a rapidly growing branch of glacial indicator tracing. The increased use of till geochemistry for mineral exploration and regional surveys has resulted in the discovery of several ore deposits and the documentation of many dispersal trains from mineralizations and distinctive bedrock units. Coker and DiLabio (1989) summarize more than one hundred mostly geochemical drift prospecting cases from North America (mainly Canada) and northern Europe (mainly Sweden and Finland), and conclude that certain specific grain size ranges and mineralogical forms contain the bulk of the metals, depending on the species of the primary metal-bearing mineral and the history of glacial comminution and weathering.

Geochemical dispersion in glacial sediments is a result of glacial deposition and postglacial processes. Bølviken and Gleeson (1979) divide therefore geochemical dispersion into two main classes: 1. syngenetic dispersion, i.e. mechanical glacial dispersion, and 2. epigenetic dispersion, i.e. chemical and mechanical dispersion after glaciation. Glacial dispersion of metals is dealt with in the article by DiLabio in this volume. Bølviken and Gleeson go on to state that downslope epigenetic dispersion pattern can be produced in glacial sediments due to metal dispersion in the groundwater, or immediately over the bedrock source due to capillary forces, biological activity or gaseous movement of volatile compounds. Sulphides and other minerals in the bedrock and soil will undergo postglacial weathering, some products of which may later be precipitated, complexed or absorbed in the soil, and this will lead in time to a concentration of metals in the soil. This is hydromorphic dispersion. Humus plays an important role in the postglacial enrichment of heavy metals.

Shilts (e.g. 1975) has emphasised that in order to use glacial sediments effectively for mineral exploration it is necessary to understand the cause of vertical compositional variability in metal values, whether sedimentological, stratigraphic or diagenetic. Any drift or stream sediment sampling programme should begin with an orientation study to determine to what extent weathering effects may mask geochemical signatures of provenance (Shilts and Kettles, this volume).

Shilts and Kettles also write that the importance of understanding contrasting compositions of different stratigraphic units is well recognized if not always adequately dealt with, in the mineral exploration literature. Diagenetic alteration of drift, on the other hand, has received little attention in the literature, and consequently it is assigned rather more emphasis in the present volume. The site described by Shilts and Kettles is near Thetford Mines, Quebec, an area of extensive drift exploration (e.g. Shilts 1976, DiLabio, this volume).

*Importance of the grain size analysed*

The grain size fraction analysed is often critical in geochemical investigations. Perhaps the most commonly used fraction is the silt+clay fraction, i.e. less than 0.063 mm, since it can easily be separated out by dry sieving, although it has disadvantages in some cases. Its coarse end normally contains large amounts of feldspars and quartz, which are 'dead' in terms of their trace metal content thus diluting the metal contents of the whole sample. This is still more evident in the fraction less than 0.15 mm, which is also used on occasion.

Sulphides and carbonates are easily affected by oxidation and leaching in the zone above the groundwater surface. They are important ore minerals and their metals are bound in finer grains, mostly to clay minerals, which have a high cation exchange capacity. Thus the anomalies caused by this hydromorphic alteration in postglacial times can perhaps be defined more precisely in some cases by analysing the clay fraction of the till (less than 0.002 mm). This has been shown on a number of occasions in Canada (e.g. Shilts 1975, 1982). The laboratory procedure for differentiating the clay fraction in a bulk sample is simple and effective, and involves repeated decanting and centrifuging. One person can handle 50 samples per day, and the method also separates out i.a. the fine gravel fraction, which can be used for till lithological analysis (Shilts, personal communication 1988, Shilts and Kettles, this volume).

The coarser silt and sand fractions should be separated out for some heavy mineral analyses, but no universal rule can be given for what is a suitable grain size when searching for gold, for instance, as its grain size naturally varies according to its original grain size in the bedrock and the secondary precipitation processes which it has undergone. The silt+clay fraction was superior in one case in Canada (Bird and Coker 1987), whereas the correlation between panned gold and instrumentally analysed gold is often very low as in the placer gold area of Finnish Lapland, for example (Saarnisto and Tamminen 1987), due to the 'nugget effect' (Clifton et al. 1969, DiLabio 1988), i.e. the fact that coarse gold in till or gravel will not show up in a chemical analysis whereas fine gold or chemically bound gold is more evenly distributed and can be detected by chemical analyses performed on small samples of 1–2 g.

*Essence of the sampling procedure*

In addition to the inportance of grain size there are a number of factors which should be considered during the sampling procedure for the purposes of drift geochemistry:
– The sample analysed may not consist of the expected material, i.e. if a percussion drill with a flow-trough bit is used, the small sample size will prevent identification of the soil type, and sorted material may be sampled instead of till, for example.
– The values given by oxidized samples are not comparable with those for unoxidized ones. The groundwater level is critical in this respect, and the use of different horizons in podsol profiles should be avoided.
– The samples for geochemical prospecting are taken from as close to the bedrock surface as possible. The lowest drill sample may contain material from the bedrock itself – weathered or unweathered – or a single boulder, so that it is the bedrock that will be analysed and not the till. Experienced field staff will recognize this source of error.
– Reworking of material. The fine fraction of the till may be reworked marine clay, for example, so that its provenance will be more or less impossible to assess, and at least its source will not be the same as that of the coarse till material even though their glacigenic dispersal may be the same. Reworking of coarser material can sometimes be recognized from the roundness of the pebbles and stones.

These factors indicate how crucial the sampling procedure is. In regional geochemical surveys the factors causing apparent variations in trace metal values may overshadow the true variations originating from lithological differences in the till. Thus the material sampled for further analysis should be genetically similar. Glaciofluvial material is not comparable with till, and different genetic varieties of till i.e. basal v. ablation or superficial tills, differ in their texture, lithology and provenance. In the case of percussion drill samples the small sample size prevents any closer genetic identification, but knowledge of the glacial geology of the target area helps in sediment analysis. Available or purposely made open sections are of great help when evaluating the nature of glacial sediments. Till statigraphy, sorted beds and layers, deviating transport directions and the preglacial weathered regolith all influence the samples to be analysed. This means that the glacial geology of the area should be worked out before embarking on a detailed sampling programme.

*Length of a geochemical anomaly*

The geochemical anomaly produced by an ore outcrop may be detectable in the till only over a very short distance, because of the limited size of the outcrop, rapid comminution of the ore minerals, especially sulphides, and mixing and leaching with the geochemically 'dead' bulk of the till matrix. On the other hand, ore boulders can be transported over long distance thus giving an impression of longer transport than for till fines. A common result when studying the transport distances of non-ore materials, is that the fine matrix represents a much wider provenance area than do the boulders, as discussed above. In practical exploration work, however, a short geochemical dispersal train should be taken as a fact, as shown in a number of examples, e.g. by DiLabio in this volume and Kauranne (1976), Kauranne et al. (1977), and Salminen (1980). The large Outokumpu copper ore deposit in eastern Finland has produced a copper anomaly in the till detectable only some hundreds of metres from the suboutcrop (Salminen and Hartikainen 1985) and yet the Kivisalmi boulder, several cubic metres in size, which lead to the discovery of the ore was found some 50 km southeast of Outokumpu, where, according to the current interpretation, it had been transported by two separate glacial flows (Kujansuu and Nenonen 1987).

*Drift geochemistry in permafrost areas*

Permafrost areas form a special case as regards drift geochemistry. Their importance is ingreasing as exploration expands to the vast permanently frozen areas of North America and Sibiria. Shilts (1973 b) investigated esker and till material in the permafrost area of Keewatin in Canada and his geochemical results point to a clear contrast between till and esker chemistry.

In 'mud boils', also termed 'non-sorted circles', which are formed on till surfaces by frost action, the normal soil horizons are quickly disrupted. Continuous recycling of sand and coarse fractions of till takes place, and repeated exposure to oxidation at the surface causes the labile minerals (sulphides, carbonates etc.) to be removed from the active zone sediment. Trace-metal analysis therefore merely reflects the metal concentrations of the stable minerals, primarily silicates. It is probable that humic acids and acids produced as by-products of sulphide weathering also contribute to the complete destruction of labile minerals within the active zone.

The relative stability of the seasonally thawing zone on eskers as compared with till surfaces accounts for the strong contrasts in background values and trace element concentrations in the fine fractions of both sediment types. The background concentrations of trace metals in the less than 0.063 mm fraction from eskers in the Keewatin target area are roughly six to ten times those for adjacent tills, due to the predominance of secondary weathering products in the esker samples. The same till fraction can also reflect anomalous areas, but in addition to the background values, the contrasts are much lower than for eskers. Thus frost activity or freeze-thaw processes will add to the basic geochemical contrast between till and esker material.

GLACIAL DISPERSAL TRAINS

Shilts (1976) writes:'Glaciers appear to disperse material in the form of a negative exponential curve, with the concentration of elements, minerals and rocks reaching a peak in till at or close to the source, followed by an exponential decline in the direction of transport'.

One manifestation of glacial indicator transport and deposition is a dispersal train. DiLabio gives details of various dispersal trains, both clastic and geochemical, and Bouchard and Salonen consider surficial boulder transport in shield areas. The characteristics of dispersal trains are summarized by DiLabio as follows:

Debris eroded glacially from a distinctive source forms a dispersal train – an elongated cluster of clasts or till lens that is enriched in the distinctive component relative to the till underlying or enclosing it. Dispersal trains can be investigated with regard to their three-dimensional shapes and structure. Many trains are very thin in comparison to their length and width, so that T.W.L. ratios of the order of 1:200:1000 are common. The following characteristics can be helpful for the prospector. Firstly, dispersal trains are much larger than their bedrock sources, making them easier targets to find. Secondly, they are usually straight and oriented parallel to the direction of ice flow, so they can be followed up-ice to the source. Thirdly, they climb gently (1° to 3°) in the down-ice direction. Fourthly, they are often very thin, which forces geologists to sample sections and drill cores at short intervals. Often long samples are needed in vertical sections to ensure discovery of the ore-bearing till. This is also discussed in the article by Puranen.

In a horizontal direction, boulder fans, the classic example of dispersal trains, characteristically coincide in orientation with the last or most pronounced direction of glacial flow, occupying a sector of approximately 10° according to Salonen (1986), although this angle may be considerably wider if the fan is a product of a number of ice flows of varying orientation. Thus fans opening up to angles of as much as 90° have been described in eastern and northern Finland in particular. Most of the 450 boulders fans identified in Finland are 1–5 km in length, although great variation is shown in this respect (Salonen 1986, 1987). Local boulder fans are only some tens or hundreds of metres in length, and the longest examples described to date measure 50–100 km, but individual boulders have in places travelled several hundred kilometres from their point of origin. This is especially the case at the marginal areas of continental ice sheets.

Salonen (1986, 1987) concludes that the ice flow in glacial lobes leads to highly diversified characteristics of dispersal, namely short, fan-shaped patterns in the head and terminal areas of the lobes and long, ribbon-shaped ones in the central parts. The dispersal patterns are often complicated by the recycling of boulders transported by earlier lobe flows (see also Hirvas and Nenonen, this volume: Complex transport).

FROM REGIONAL SURVEY TO SUCCESSFUL INDICATOR TRACING

Distinct bedrock units although not economically valuable, are useful in regional indicator tracing when working out the directions and distances of glacial transport. An example from Finland will illustrate this (Saarnisto and Taipale 1985, see also Hirvas and Nenonen, this volume, Fig. 10). It comes from the area of the Archaean Kuhmo greenstone belt in eastern Finland, where the glacial stratigraphy, direction and distances of glacial transport together with the till lithology and geochemistry were worked out on a regional scale. The principal flow directions of the continental glacier cross the 3–5 km broad greenstone belt almost at right angles over most of its length. The peak concentrations of coarser greenstone material in the till were found on top of the belt itself, the values decreasing to a half within only 0.5 km in the direction of ice movement. The boulders were more local in origin than the 2–6 cm diameter pebbles, which still contributed small amounts to the till at a distance of 15 km from the greenstones.

The greenstone belt is similarly reflected in the trace metal content of the silt+clay fraction of the till, which rapidly doubles on top of the belt, and the concentration of Ni even increases four-fold. The highest values are often found on top of the greenstone belt, although occasionally also on the distal (down-glacier) side. The maximum values normally decrease to a half again within 0.5 km. The Cu, Mn, Pb, Zn and Fe values for the background (the granitoids west of the belt) are not, however, attained until a distance of 20 km from the greenstones in the direction of ice movement is reached. Here Ni and Co still remain higher than the background content.

Broadly speaking, the lithology of the coarse till fractions and the fine fraction geochemistry seem to reflect the presence of the greenstones in a similar way, although the trace-metal values are low as compared with the high values in greenstones. The contrasts between the greenstones and the surrounding granitoids are so clear, however, that even a small proportion of greenstones can raise the metal content of the till. The peak values are very local, a fact, which should be bept in mind during geochemical exploration.

Once the glacial stratigraphy, principal transport directions and most probable transport distances had been worked out in a regional scale, glacial indicator tracing was successfully employed in the southern part of the Kuhmo greenstone belt to locate a silver – zinc – lead mineralization, as described below.

Surface boulders of felsic metavolcanites containing zinc and lead were found on the shores of Lake Taivaljärvi in Sotkamo. This area was considered a likely one for zinc-lead mineralization on the basis of pure geological research because it is situated close to an ancient volcanic centre. In order to find the source of the boulders, till samples were taken with a light-weigth percussion drill from as close to the bedrock as possible at 50 metre intervals in the direction of glacial flow. The average till thickness was 5 metres.

600 metres upglacier from the boulders a Zn-Pb mineralization was discovered under 6 metres of till cover by analysis of the till chemistry. Two sampling points hit the mineralized zone, which was about 70 metres wide at the survey transect. Later investigations have shown that the first boulders found were also closest to the ore suboutcrop, i.e. they were proximal boulders (see Hirvas and Nenonen, this volume). It was also soon discovered that the mineralization is rich in silver, as were the first boulders, silver being the most valuable metal present, accounting for more than 70% of the total metal value of the ore (Kopperoinen and Tuokko 1988).

The Taivaljärvi mineralization is not recognizable on geophysical aeromaps, since an iron formation situated 500 m upglacier overshadows other magnetic anomalies within the area. In fact, there is a slight IP anomaly marking the mineralization, which only becomes evident when the deposit has first been recognized by other methods, i.e. genetic and lithological bedrock studies, boulder tracing and till geochemistry. An extensive diamond drilling programme was started at the site together with surveys of the till, peat and lake sediment geochemistry. The mineralization does not show up in the lake sediments or in the peat on top of it, but its presence is evident in the till, with lead and silver in particular serving as good indicators, whereas zinc is more complicated, obviously because of its mobility in hydromorphic processes. The lead anomaly is situated just on top of the mineralization, where concentrations at least 10 times higher than the background values of 10 ppm are common. The silver anomaly is also a clear one the highest values being situated only 30 to 40 metres down-glacier from the ore outcrop. In fact the bottom-most till samples on top of the ore outcrop contain as much silver as the ore itself (in the order of 100 ppm). The Taivaljärvi indicator train closely resembles in its vertical dimensions and form those presented by DiLabio in this volume.

Several million tons of mine-grade deposit have been investigated so far, and the construction of a mining tunnel was started in September 1988.

REFERENCES

Baker, C.L., 1982. Report on the sedimentology and provenance of sediments in eskers in the Kirkland Lake Area, and on finding of kimberlite float in Gauthier Township. In: J. Wood, O.L. White, R.B. Barlow and A.C. Colvine (eds.), Summary of Field Work. Geological Survey of Ontario, Miscellaneous Paper 106: 125–127.

Bird, D.J. and W.B. Coker, 1987. Quaternary stratigraphy and geochemistry at the Owl Creek Gold Mine, Timmins, Ontario, Canada. In: R.G. Garret (ed.), Geochemical Exploration 1985. Part 1. Journal of Geochemical Exploration 28: 267–284.

Boulton, G.S., 1975. Processes and patterns of subglacial sedimentation, a theoretical approach. In: A.E. Wright and F. Moseley (eds.), Ice Ages: Ancient and Modern. Seel House, Liverpool: 7–42.

Bølviken, B. and C.F. Gleeson, 1979. Focus on the use of soils in geochemical exploration in glaciated terrane. In: P.J. Hood (ed.), Geophysics and Geochemistry in the Search for Metallic Ores. Geological Survey of Canada, Economic Geology Report 31: 295–326.

Clifton, H., R. Hunter, F. Swenson and R. Phillips, 1969. Sample size and meaningful gold analysis. U.S. Geological Survey, Professional Paper 625-C, 17 pp.

Coker, W.B. and R.N.W. Dilabio, 1989. Geochemical exploration in glaciated terrain: geochemical responses. In G.D. Garland (ed.), Proceedings of Exploration '87, Ontario Geological Survey, Special Volume 3:336–383.

DiLabio, R.N.W., 1988. Residence sites of gold, PGE, and rare lithophile elements in till. In: D.R. MacDonald and K.A. Mills (eds.), Prospecting in Areas of Glaciated Terrain – 1988. The Canadian Institute of Mining and Metallurgy: 121–140.

Dreimanis, A. and J.J. Vagners, 1971. Bimodal distribution of rock and mineral fragments in basal tills. In: R.P. Goldthwait (ed.), Till: A Symposium. Ohio State University Press: 237–250.

Dreimanis, A. and J.J. Vagners, 1972. The effect of lithology upon texture of till. In: E. Yatsu and A. Falconer (eds.), Research methods in Pleistocene geomorphology. Geo Abstract Ltd: 66–82.

Evenson, E.B. and J.M. Clinch, 1987. Debris transport mechanisms at active alpine glacier margins: Alaskan case studies. In: R. Kujansuu and M. Saarnisto (eds.), INQUA Till Symposium, Finland 1985. Geological Survey of Finland, Special Paper 3: 111–136.

Evenson, E.B., T.A. Pasquini, R.A. Stewart and G. Stephens, 1979. Systematic Provenance Investigations in Areas of Alpine Glaciation: Applications to Glacial Geology and Mineral Exploration. In: Ch. Schlüchter (ed.), Moraines and Varves. A.A. Balkema, Rotterdam: 25–42.

Flint, R.F., 1971. Glacial and Quaternary Geology. John Wiley and Sons, New York, 892 pp.

Gillberg, G., 1965. Till distribution and ice movements on the northern slopes of the South Swedish Highland. Geologiska Föreningen i Stockholm Förhandlingar 86: 433–484.

Gross, D.L. and S.R. Moran, 1971. Grain size and mineralogical gradations within tills of the Allegheny Plateau. In: R.P. Goldthwait (ed.), Till: A Symposium. Ohio State University Press: 251–274.

Haldorsen, S., 1977. The petrography of tills – a study from Ringsaker, south-eastern Norway. Norges Geologiska Undersøgelse 336, 36 pp.

Harrison, P.W., 1960. Original bedrock composition of Wisconsin till in central Indiana. Journal of Sedimentary Petrology 30: 432–446.

Jørgensen, P., 1977. Some properties of Norwegian tills. Boreas 6: 149–157.

Kauranne, K., R. Salminen and M. Äyräs, 1977. Problems of Geochemical Contrast in Finnish Soils. In: M.J. Jones (ed.), Prospecting in Areas of Glaciated Terrain (1977) Inst. of Mining and Metallurgy, London: 34–44.

Kopperoinen, T. and I. Tuokko, 1988. The Ala-Luoma and Taivaljärvi Zn-Pb-Ag-Au deposits, eastern Finland. In: E. Marttila (ed.), Archaean Geology of the Fennoscandian Shield. Geological Survey of Finland, Special Paper 4: 131–144.

Kujansuu, R., 1987. Opening address. In: R. Kujansuu and M. Saarnisto (eds.), INQUA Till Symposium, Finland 1985. Geological Survey of Finland, Special Paper 3: 9–10.

Kujansuu, R. and K. Nenonen, 1987. Till stratigraphy and ice-flow directions in North Karelia. Geological Survey of Finland, Special Paper 1: 59–66.

Lee, H.A., 1965. Investigations of eskers for mineral exploration. Geological Survey of Canada, Paper 65–14, 20 pp.

Lindén, A., 1975. Till petrographic studies in an Archaean bedrock area in southern central Sweden. Striae 1, Uppsala, 57 pp.

Martin, D.P. and Eng, M., 1985. Esker prospecting over the Duluth complex in northeastern Minnesota. Minnesota Department of Natural Resources, Division of Minerals, Report 246, 27 pp.

Milthers, V., 1909. Scandinavian indicator boulders in the Quaternary deposits; extension and distribution. Danmarks geologiske Undersøgelse, Ser. 2, No. 23, 154 pp.

Minell, H., 1978. Glaciological interpretations of boulder trains for the purpose of prospecting in till. Sveriges Geologiska Undersökning, Ser. C 743, 51 pp.

Nevalainen, R., 1989. Lithology of fine till fraction in the Kuhmo greenstone belt area, eastern Finland. Geological Survey of Finland. Special Paper 7: 59–65.

Perttunen, M., 1977. The lithological relation between till and bedrock in the region of Hämeenlinna, Southern Finland. Geological Survey of Finland, Bulletin 291, 68 pp.

Prest, V.K. and E. Nielsen, 1987. The Laurentide Ice sheet and long–distance transport In: R. Kujansuu and M. Saarnisto (eds.), INQUA Till Symposium, Finland 1985. Geological Survey of Finland, Special Paper 3: 91–101.

Puranen, R., 1988. Modelling of glacial transport of basal tills in Finland. Geological Survey of Finland, Report of Investigation 81: 36 pp.

Saarnisto, M. and K. Taipale, 1985. Lithology and trace-metal content in till in the Kuhmo granite-greenstone terrain, eastern Finland. Journal of Geochemical Exploration 24: 317–336.

Saarnisto, M. and E. Tamminen, 1987. Placer gold in Finnish Lapland. In: R. Kujansuu and M. Saarnisto (eds.), INQUA Till Symposium, Finland 1985. Geological Survey of Finland, Special Paper 3: 181–194.

Salminen, R., 1980. On the geochemistry of copper in the Quaternary deposits in the Kiihtelysvaara area, North Karelia, Finland. Geological Survey of Finland, Bulletin 309, 48 pp.

Salminen, R. and A. Hartikainen, 1985. Glacial transport of till and its influence on interpretation of geochemical results in North Karelia, Finland. Geological Survey of Finland, Bulletin 335, 48 pp.

Salonen, V-P., 1986. Glacial transport distance distribution of surface boulders in Finland. Geological Survey of Finland, Bulletin 338, 57 pp.

Salonen, V-P., 1987. Observation on boulder transport in Finaland. In: R. Kujansuu and M. Saarnisto (eds.), INQUA Till Symposium, Finland 1985. Geological Survey of Finland, Special Paper 3: 103–110.

Saltikoff, B., 1984. Boulder tracing and mineral indication data bank in Finland. In: Prospecting in Areas of Glaciated Terrain (1984). Institution of Mining and Metallurgy, London: 179–191.

Shilts, W.W., 1971. Till studies and their application to regional drift prospecting. Canadian Mining Journal 4 (1971): 45–50.

Shilts, W.W., 1973a. Glacial dispersal of rocks, minerals and trace elements in Wisconsinan till, southeastern Quebec, Canada. In: R.B. Black, R.P. Goldthwait and H.B. Willman (eds.), The Wisconsinan Stage. Geological Society of America, Memoir 136: 189–219.

Shilts, W.W., 1973b. Drift prospecting; Geochemistry of eskers and till in permanently frozen terrain: District of Keewatin, Northwest Territories. Geological Survey of Canada, Paper 72–45, 34 pp.

Shilts, W.W., 1975. Principles of geochemical exploration for sulphide deposits using shallow samples of glacial drift. Canadian Institute of Mining and Metallurgy, Bulletin 68(757): 73–80.

Shilts, W.W., 1976. Glacial till and mineral exploration. In: R.F. Legget (ed.), Glacial Till: An Interdisciplinary Study. Royal Society of Canada, Special Publication 12: 205–224.

Shilts, W.W., 1982. Glacial Dispersal – Principles and Practical Applications. Geoscience Canada 9(1): 42–47.

Shilts, W.W., 1984. Till Geochemistry in Finland and Canada. Journal of Geochemical Exploration 21: 95–117.

Svantesson, S.-I., 1976. Granulometric and petrographic studies of till in the cambrosilurian area of Gotland, Sweden, and studies of the ice recession in northern Gotland. Striae 2, Uppsala, 79 pp.

Taipale, K., R. Nevalainen and M. Saarnisto, 1986. Silicate analyses and normative composition of the fine fraction of till: Examples from eastern Finland. Journal of Sedimentary Petrology 56: 370–378.

Virkkala, K., 1971. On the lithology and provenance of the till of a gabbro area in Finland. VIII Congrès INQUA, Paris 1969. Etudes sur le Quaternaire dans le Monde 2: 711-714.

# CHAPTER 2

# Modelling of glacial transport of tills

RISTO PURANEN
*Geological Survey of Finland, Espoo*

## INTRODUCTION

The dispersal of till material by glacial ice has been studied during the past hundred years or more by comparison of the lithologic, mineralogic and chemical composition of tills and their source rocks. These studies have led to qualitative understanding of the transport process of till material as well as to estimates of the transport distances in different areas and for different grain sizes of till. The studies have also shown that the amount and distribution of source material in till depend on the size and rock type of the source area, on the thickness of the till sheet and on whether the samples were taken from the surface or bottom of the till bed.

The elaboration of geologic data into a transport model has been limited to fitting mathematical curves to the data points, showing that the abundance of source material in till decreases on the distal side of bedrock sources. Lines and exponential curves have been fitted to these declining abundance values, and the parameters of such curves have been compared for various till fractions, source rock types and research areas. The curve most commonly fitted is probably that of a negative exponential function known in Quaternary geology as Krumbein's (1937) exponential curve. From the curve such parameters as the 'transport half-distance' (Gillberg, 1965) or comminution index 'a' (Bouchard et al., 1984) are determined.

In this paper the glacial geologic data from Finland are summarized into statements that must be satisfied by a glacial transport model. A transport model is then derived, based on glaciologic concepts, and it is tested against the geologic data. The ranges of the glaciologic parameters are established, and the behaviour of the model is examined accordingly for realistic parameter values. The transport distances and times predicted by the model are compared with those determined geologically. The relations between anomalous till layers and their source areas are described and the significance of sampling to the interpretation of transport profiles is emphasized. Some examples are given to demonstrate the application of the model to the geochemical sampling design and interpretation. Finally, the boulder transport and different transport conditions are discussed.

## GEOLOGIC CONSTRAINTS

The geologic observations of glacial transport in Finland, as reviewed by Puranen (1988), can be summarized into the following *statements*:

1. The till cover rests on bedrock, which is topographically level when compared with the height of continental ice, but uneven compared with the thickness of till cover.

2. The till cover is usually composed of 1–2 basal till beds, whose thickness is generally 1–3 m. In many places the basal till is overlain by a layer of ablation till, which is often less than 1 m thick (Hirvas and Nenonen,1987).

3. The bulk of the till has been transported less than 10 km from its source. After a contact source, the coarse material in basal till is renewed over a distance of 1–10 km (Hellaakoski, 1930; Virkkala, 1971; Perttunen, 1977; Salminen, 1980; Salonen, 1986).

4. The transport distance of basal till is shorter in the lower parts of the till layer than in the upper parts (Sauramo, 1924).

5. The transport distance of a thin till layer is shorter than that of a thick one (Hirvas et al., 1977).

6. The basal till is renewed more rapidly above brittle (felsic) rock types than above resistant (mafic) rocks (Puranen, 1988).

7. The proportion of dike material in basal till reaches its maximum close to the distal contact of the dike source, whereafter it decreases asymptotically towards zero (Shilts, 1976; Salminen and Hartikainen, 1985).

8. The maximum proportion of dike material in till increases with the width of dike (Peltoniemi, 1985).

9. The transport distance of dike material is shorter in the lower parts of the till bed than in the upper parts (Nurmi, 1976; Salminen and Hartikainen, 1985).

10. The transport distance of till at the ice divide in Lapland is shorter than in southern Finland (Pulkkinen et al.,1980; Lehmuspelto, 1987).

The above observations were mostly obtained from investigation lines running in the glacial flow direction approximately perpendicularly over dike or contact sources.


OUTLINING THE MODEL

A framework is outlined for a transport model, whose simplicity is based on the use of averages to describe the complex geologic reality. The model grows out of transport profiles of till that repeat themselves in a similar fashion from one area to the next. Modelling will be started with a 2-dimensional case, and the glacier substratum is assumed to be level in accordance with *statement* 1. The model deals at first with a case, in which only one basal till layer is formed on the bedrock. The expansion of the model to multi-layer cases is presented by Puranen (1988).

*Statements* 2 and 3 limit the vertical dimension of transport modelling to a couple of meters and the horizontal dimension to about 10 km. In glaciology, physical models have been presented both for growth of large ice sheets (dimension > 1000 km), and for erosion caused by small grains (dimension < 1 m). Models with intermediate dimensions of about 10 km have not been presented, as pointed out by geologists in their inspiring dialogue with glaciologists (Hallet, 1981). Since the following model has a horizontal dimension of about 10 km, a geologically reasonable 1 % resolution is obtained with the model even if the distance steps are raised to 100 m. The effect of small roughness and heterogeneity of the source bedrock can thus largely be eliminated by averaging the model parameters over a distance of 100 m.

The model describes the material movement as a laminar flow that takes place along layers. This is compatible with the tightly packed layering often encountered in basal tills. In the model, deposition is simply treated as the end of transport, although in reality intermixing of material naturally takes place during both transport and deposition. The description of intermixing could be later added to the model as either a statistical or a physical diffusion function. On the other hand, the influence of minor intermixing can be eliminated by studying the mean abundances of longer samples. The mean values give the same result that would be obtained by complete mixing of anomalous material into the samples.

Let us first consider a hypothetic situation in which a diabase dike is located in the middle of extensive and flat granitic bedrock area. The dike is 100 m wide and trends perpendicular to the glacial flow direction. The dike is assumed to be very long, thus allowing us to consider the transport as a 2-dimensional case. Let the bedrock be overlain by a basal till layer with an average thickness of 2 m. Let us further assume that the ice sheet produced the basal till solely by dislodging, breaking and grinding the underlying bedrock material without transporting it. In other words, granitic till material would only be found over granite, and the diabase material would form a cake, 100 m wide and 2 m high, directly above the source dike.

However, we know that the crushed and grinded rocks have been transported and that they have been spread some distance downglacier from their source. If we distribute the diabase cake evenly over a distance of 10 km, the result will be a 2-cm-thick diabase layer in a 2-m-thick till. The diabase layer would represent about one percent of the total thickness of the till bed, and within the framework of typical geologic variation, it could be difficult to even notice such a compositional layer. The cake that develops above a diabase dike or a thin layer derived from a dike represent extreme cases, and the true distribution of diabase material in basal tills lies between them.

Numerous geologic (Sauramo, 1924; Drake, 1983) and geochemical studies (Halonen, 1967; Nurmi, 1976; Miller, 1984) describe the dispersal of material from dike sources as ribbons, fans, surfaces or plumes that rise from the source towards the till surface at varying angles. Sketched in Fig. 1A is an anomalous ribbon of till material rising from the dike source. The abundance profile shows how the proportion of dike material in the vertical till cross-section varies along the transport direction. The abundance begins to rise at the proximal contact of the dike, reaches its maximum at the distal contact and declines thereafter. Figure 1B illustrates in a similar fashion the case of a contact source formed by the boundary between two rock types. To produce the rising ribbons we need forces that lift the material and forces that transport it. Furthermore, the influence of the transporting forces need to be greater than that of the intermixing forces, since in most instances till fabric and anomalies have retained their regularity.

During glacial transport the till material also undergoes comminution, which transfers material from the coarse grain sizes to the finer fractions. Consequently, the abundance of coarse material from a dike source declines in till more rapidly than does that of the fines. Studies on coarse till fractions thus indicate transport distances that differ from those given by fines (Peltoniemi, 1985). If we want to determine the absolute movement of material in the glacial transport, we have to study unsieved till material. In this way we can establish the total transport of till material without the interference of comminution. By comparing the transport of unsieved material with the apparent transport of different fractions we can study the comminution process as well. The following model describes only the transport of unsieved till material, and the modelling is further restricted to the

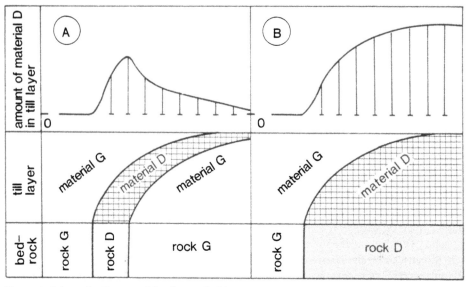

Figure 1. Schematic diagram of the dispersal of basal till material from A) dike source and B) contact source.

glacigenic (mechanic) movement of the material. The postglacial phenomena are omitted but their influence is discussed elsewhere (Puranen, 1988).

### DERIVING THE MODEL

The movement of an ice sheet can be described with the aid of creep and basal sliding. Creep is due to the gravity forces that tend to spread the glacier under its own weight towards its margins. If the base of glacier is frozen to the substratum, the creep is the only mode of movement in the glacier. Such is the situation in large areas of Greenland and Antarctica, where erosion is weak due to the lack of movement of the cold-based ice sheets. Erosion takes place only if the glacier sole moves in relation to its bed, as in warm-based glaciers. It is not easy to observe the basal sliding under the glacier, and so glaciologists have developed complex mathematical sliding theories (regelation, enhanced plastic flow) to describe the phenomenon (see Paterson, 1981).

The creep velocity depends on the mass balance, thickness, surface slope and temperature of the glacier, as well as on the concentration of debris in the basal ice layers. The variation of horizontal creep velocity $U(z)$ as a function of distance z from the base of the glacier has been studied with the aid of borehole measurements. The velocity is highest on the surface of the glacier and decreases towards the bottom of the glacier. The following model assumes that in the basal parts of the glacier the velocity U increases linearly with the distance z, or $U(z) = k \cdot z$, which is a reasonable assumption according to observations. The proportionality constant k is the (engineering) strain rate defined by $dU/dz$. The basal sliding velocity of the glacier depends on the temperature and water conditions at the glacier base, the friction between the glacier base and the bed, and the

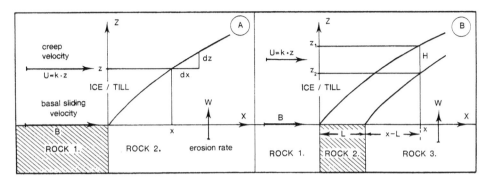

Figure 2. Symbols used in the transport model of basal tills in the case of A) contact source and B) dike source.

roughness of the bed. In the following, the velocity of basal sliding is assumed to be constant, and it is denoted with the symbol B. For the total velocity V of glacier motion we then obtain $V(z) = B + U(z) = B + kz$.

The erosion rate of the glacier bed depends on the properties of the glacier and the bed (see Drewry, 1986). The rate of erosion increases with the basal sliding velocity and the thickness of the glacier (effective normal pressure on the bed). Erosion rate is also dependent on the coarseness and hardness of the rock or soil types of the substratum. Another significant factor controlling the rate of erosion is the relative difference in hardness of the rock material in the glacier base and the bedrock. The erosion rate, which is assumed to be constant in the following transport model, is denoted with the symbol W. To be more accurate, the parameter W refers to the rate at which the thickness of the basal till increases, which is slightly higher than the erosion rate of the bedrock, because the bulk density of till is lower than the density of rock.

Let us consider the situation depicted in Figure 2A, in which the glacier is flowing on an even substratum from left to right. The glacier bed is composed of two lithologic areas, whose contact is located in the middle of the figure at the origin of the coordinate system. The transport distance increases along the X-axis to the right, and the distance from the bottom of the glacier increases along the Z-axis upwards. Also drawn in the figure is a curve starting at origin and describing the transport path of a single rock particle dislodged from bedrock. Let us examine a differential movement of the particle along the curve at a distance z from the glacier base. During a time inverval, dt, the particle moves away from the source a horizontal distance dx and a vertical distance dz. Because the glacier velocity at depth z is $V = B + kz$, we get for the horizontal displacement

$$dx = (B + kz)dt \qquad (1)$$

During time dt, a new layer of dislodged rock particles is formed under the glacier. This layer is incorporated into the base of the glacier, which forces the earlier eroded grains upwards from the base. The rock particle under consideration is thus moved upwards for a distance

$$dz = Wdt \qquad (2)$$

From equations (1) and (2) we obtain

$$dx = (1/W)(B + kz)dz \tag{3}$$

By integrating equation (3) we get for the path of the grain:

$$x = (B/W)z + \tfrac{1}{2}(k/W)z^2 \tag{4}$$

or

$$x = z(B + \tfrac{1}{2}U)/W \tag{5}$$

Equation (4) shows that the transport path is completely defined when ratios B/W and k/W are known. Inversely, from the observed path it is possible to determine the above ratios but not the parameters separately. Hence, the value of one of the parameters B, W or k can be set freely, which then fixes the values of the others.

According to equation (5), the shape of the transport path depends on whether basal sliding (B) or creep (U) is dominating. The path approximates a straight line when basal sliding dominates and a parabola when creep dominates the glacial movement. When equation (4) is solved for the depth z, we get

$$z = f(x) = -S + \sqrt{(S^2 + 2Wx/k)} \tag{6}$$

where   $S = B/k$ $\tag{7}$

Equation (4) describes the transport path of one rock particle dislodged from bedrock. Since the particle derives from the contact of two rock types, the curve shows the boundary between the grains derived from rock 1 and rock 2, respectively. According to the model, the till above the boundary curve is totally composed of rock 1 and the till below the curve contains 100% of rock 2. In reality, the boundary is less sharp as a result of mixing processes, which, however, are not treated in the present study.

The contact situation described can also be interpreted as a dike case, in which rock 2 forms a dike with semi-infinite width. The case of a narrow dike is formed by two adjacent contacts that produce two successive contact curves as shown in Figure 2B. These curves distinguish the dike material in till from the material derived from other rocks. Let us consider the situation at a distance x from the proximal contact of a dike that has a width L. The material from this contact is found in the till at the height $z_1 = f(x)$ according to equation (6). The distance from the observation point to the distal contact is $x - L$, and material deriving from this contact is at the height $z_2 = f(x - L)$. Hence, all the till material from the depth interval $z_1 - z_2$ originates from the dike. Since $z_2 = 0$ when $x \le$ L the height of the till column H at the point x is

$$\begin{aligned} H &= z_1 &\text{when } x \le L \\ H &= z_1 - z_2 &\text{when } x > L \end{aligned} \tag{8}$$

where $z_1$ and $z_2$ are obtained from equation (6) by giving values x and $x - L$ to the transport distances.

When the width L of the dike increases, x–L is reduced as is also $z_2$, with the consequence that $H = z_1 - z_2$ increases. The absolute height of the till column, and hence the abundance of the dike material at point x, are proportional to the dike width at low values of L. To be accurate, equation (8) holds only when the erosion rate W has the same values for the country rock and the dike. If the erosion rates differ, the transport paths of the particles deriving from the dike and the country rock intersect, causing intermixing in till, particularly around the contacts. In general, local variations in the erosion rate constitute a disturbing factor in the model. The disturbance can be reduced by increasing the cell size in modelling, which stabilizes the erosion rate through averaging over larger units.

MODEL PARAMETERS

According to Paterson (1981) and Drewry (1986) the following ranges for parameters W, B and k can be suggested:

$$W = (3–300)\cdot 10^{-12}\ \text{m/s} = 0.1–10\ \text{mm/a}$$
$$B = (3– 500)\cdot 10^{-8}\ \text{m/s} = 1–150\ \text{m/a} \qquad\qquad (A)$$
$$k = (0.3–30)\cdot 10^{-8}\ \text{s}^{-1}$$

The parameters show several kinds of intercorrelations. For instance, the erosion rate W increases with the basal sliding velocity B of the glacier. The basal sliding as well as the creep velocity U both increase as the temperature of the glacier rises. When the concentration of debris at the base of the glacier increases, the strain rate k is reduced. At the same time the friction between the glacier base and the bed increases, causing a decrease in the basal sliding velocity. We can thus generalize that the values of parameters W, B and k all increase or decrease together with each other.

Since the basal sliding velocity B and the creep velocity U are in the numerator of equation (5) and the erosion rate W is in the denominator, the positive correlation between the parameters tends to stabilize the transport estimates. Let us examine the model behaviour for different sets of parameter values. The following pairs of values W = 1 mm/a and B = 10 m/a, W = 4 mm/a and B = 20 m/a, W = 36 mm/a and B = 250 m/a, have been measured on present-day glaciers of Iceland (Boulton, 1974). By introducing z = 1 m, the average strain rate $k = 3\cdot 10^{-8}\ \text{s}^{-1}$ and the above pairs of parameter values into equation (5) we get the estimates 10.5, 5.1 and 7.0 km for the transport distance. All the calculated values are in harmony with the range of renewal distances (1–10 km) that was geologically established (*statement* 3). The calculated transport estimates are reduced only by one half although the parameter values are increased more than tenfold. This demonstrates well the tendency of the model to stabilize transport distances.

Let us now consider the model parameters in terms of time. As starting values, we shall use the pair W = 4 mm/a and B = 20 m/a. At this erosion rate a till cover averaging 2 meters in thickness would be formed within 500 years, provided the process is continuous. During that time the material would have moved 10 000 metres as a result of basal sliding of the glacier. It has been estimated that the last Scandinavian Ice Sheet existed about 20 000-30 000 years (Paterson, 1981; Boulton, 1984), of which 10 000 years could have elapsed before the ice sheet had grown to its full dimensions. Against this background the above time interval of 500 years appears short, implying that the W-value used may be too high. Longer times, e.g. 10 000 years, are obtained for the model transport process by reducing the erosion rate to W = 0.2 mm/a. This value is also within the range of the realistic values (A). When we combine a lower velocity of basal sliding B = 1 m/a with the longer time, the till material still has moved 10 000 meters owing to basal sliding. The lower B-value would be in the range of values observed below cold-based glaciers (Echelmeyer and Zhongziang, 1987).

Of the pairs I: W = 4 mm/a, B = 20 m/a and II: W = 0.2 mm/a, B = 1 m/a, the latter is the more probable in terms of time. If the glacial activity slows down in parts of the ice sheet or the basal ice stagnates, the glaciation time and the transport time may differ from each other. Then the realistic parameter values would be between those of pairs I and II. Refining the parameter limits can be continued by examining the distribution of basal till material. Both the renewal of till material after a contact source and the decrease of dike source material in the till take place in an asymptotic manner (*statement* 7). The second power term in the model equation (4), which describes creep, gives the transport paths some asymptotic character. Hence, the correct form of transport paths can be obtained only when the basal sliding and creep velocities are approximately equal. At the mean strain rate $k = 3 \cdot 10^{-8}$ s$^{-1}$, we get $U = k \cdot z = 6 \cdot 10^{-8}$ m/s = 2 m/a for the creep velocity at depth $z = 2$ m. Therefore, model curves with the right form can be expected only if the basal sliding velocities stay below 1–10 m/a. This line of approach also suggests that the parameter pair II is closer to the true values than pair I.

When deriving the model it was assumed that the material eroded at the glacier base is incorporated into the base and thus moved along with the creep of ice. However, the model can also be used if the eroded material is kept under the glacier and is there moved by deformation. In principle, the model only requires that the erosion takes place at the interface between a solid substratum and the lowermost layer of dislodged material. In practice, the locus of erosion and deformation should have an effect on the model parameters. Different parameter values are expected for situations in which the dislodged material is within the glacier as basal debris, or under a warm-based glacier as water-saturated till or under a cold-based glacier as ice-laden drift. Deformation of till under various glaciers has been described by Boulton (1979), Alley et al. (1986) and Echelmeyer and Zhongxiang (1987).

The relative importance of basal sliding and creep in transport process can vary during one glaciation cycle, which may lead to the deposition of different till lithofacies with varying transport characteristics (see Bouchard and Salonen, this volume). This case can be treated with the model by dividing the transport time into shorter periods characterized by constant parameter values. Since both the model and the observations can be considered as first-order approximations of reality, there are no grounds for further analyses of the parameters. In the following numerical examples the model parameters will be given the following set of values (P):

$$
\begin{aligned}
W &= 0.4 \text{ mm/a} \\
B &= 1.0 \text{ m/a} \\
k &= 3 \cdot 10^{-8} \text{ s}^{-1}
\end{aligned}
\qquad \text{(P)}
$$

At these values the transport distance x = 3.7 km is obtained at height z = 1 m above the bottom of till bed. In other words, the lowermost half of a 2-m-thick till layer is renewed after a transport of 3.7 km in accordance with *statement* 3. In fact, the transport distances calculated by the model are compatible with geologic observations, when in the set (P) the erosion rate, for example, is kept within 1.5–0.15 mm/a and the other parameters are constant.

TESTING THE MODEL

After the treatment of model parameters and transport distances (*statement* 3), we shall examine whether the model behaviour is in accordance with the other geologic constraints. The equation for the transport path written in form (5) shows how various factors affect the shape of the path. According to the contact case equation $x = z \cdot (B + \frac{1}{2}U)/W$, the transport distance x of a particle should increase with its distance z from the bottom of the till layer. Therefore, the transport and renewal distances of till should be shorter in the lower parts of a till layer than in its upper parts (*statements* 4 and 9). In other words, the renewal distance of a thin till cover should be shorter than that of a thick one (*statement* 5) under similar conditions.

Also, according to equation (5), the transport distance x is inversely proportional to the rate W of glacial erosion. At high erosion rates (brittle substratum) a till grain rises, for example, two meters from the substratum after a shorter transport distance than at a lower rate (resistant substratum). Consequently, the basal till composition should be renewed more rapidly (after shorter distances) when the glacier flows into an area that is composed of weak rock or sediment types (*statement* 6). Still according to equation (5), the transport distance of basal till is directly proportional to the velocities B and U of glacier movement. Since the flow velocities are lower in the central parts of the glacier than at its margins, the short transport distances at the ice divide in Lapland as compared with those in southern Finland (*statement* 10) have a natural explanation in the model.

The case of a dike source is next considered with the aid of equation (8). The equation allows us to calculate the height H (or amount) of the material transported to the distance x from the proximal contact of the dike. Let the width of the narrow dike be L, and the glacier parameters be expressed with symbol S = B/k. To facilitate the considerations the dike case equations are repeated below

$$H = z_1 \qquad \text{when } x \leq L$$
$$H = z_1 - z_2 \qquad \text{when } x > L$$

where $\qquad\qquad\qquad\qquad\qquad\qquad\qquad\qquad\qquad\qquad\qquad$ (8b)

$$z_1 = -S + \sqrt{(S^2 + 2Wx/k)}$$
$$z_2 = -S + \sqrt{(S^2 + 2W(x-L)/k)}$$

With increasing transport distance, the curves $z_1$ and $z_2$ approach each other continuously, as x becomes increasingly larger than L. The difference $H = z_1 - z_2$ is greatest at point $x = L$, where $z_2 = 0$. In other words, the amount of dike material in the till reaches its maximum at the distal contact of the dike, after which the proportion gradually decreases with increasing transport distance (*statement* 7). The maximum amount of dike material in the till is obtained by equation

$$H = -S + \sqrt{(S^2 + 2WL/k)} \qquad\qquad\qquad\qquad\qquad\qquad (9)$$

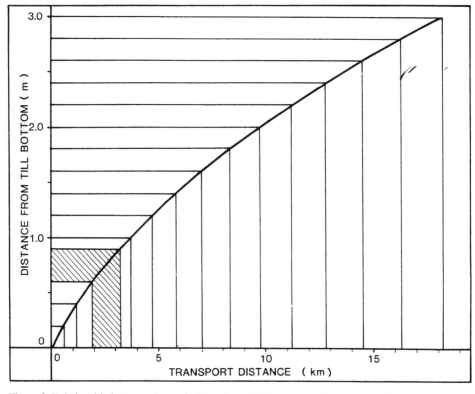

Figure 3. Relationship between the vertical location of till material and its transport distance calculated from equation (4) with parameters (P).

The value of maximum H increases with the width of the dike in accordance with the observations (*statement* 8). Since the maximum depends on the width L, the half-distance estimates based on the maximum value also depend on the dike width. When the equation (9) is solved for the dike width L, we get

$$L = (BH + \tfrac{1}{2}kH^2)/W \tag{10}$$

This equation allows the estimation of source width L from the maximum amount H of dike material, if the glacier parameters k, B and W are known or assumed.

MODEL PREDICTIONS

The model depicts the theoretic transport path of individual till particles. Although it provides a conceptual framework which can facilitate the planning of till sampling and the interpretation of results, it cannot describe in detail the complex reality. For instance, the influence of the bedrock topography is ignored in the model. In the following, a few cases will be treated with the aid of the model to illustrate its restrictions and its use for predictions. In the calculations, the glacier and its substratum are characterized by the parameter set (P). The computation results are shown as various curves, which might prove to be helpful also in practical interpretations.

In Fig. 3 the relationship between the vertical position of a till grain and the location of its source is presented on the basis of equation (4). If the grain is located, for example, at a height of 1, 2 or 3 meters from the base of the till layer, its source should be found at a distance of about 4, 10 or 18 kilometers, respectively. If again the till layer is within the height interval of 60–90 cm, its source area is expected at approximately 2–3 km from the observation site. According to the model, the compositional layers within the till material are suprisingly thin even when derived from sources that are 1 km wide. This feature of the model is illustrated in Fig. 4. If the source is, say, 100 m wide, the layer it produces is merely 2–3 cm thick, which could well go unobserved in sampling. Widening the source to 800 m increases the thickness of till material produced only to 20–30 cm. A rule of thumb can easily be derived for the height estimates of till production. If a 2-m-thick till cover is completely renewed over a distance of about 10 km (*statement* 3), then a source, 10 km wide, has produced at its distal contact a 2 meters layer of till. Correspondingly, it can be deduced that a source, 1 km wide, generates a layer thickness of 20 cm, and a source, 100 m wide, a thickness of 2 cm (see Fig. 4).

In exploration work the target zones of bedrock are often only tens of meters wide. From the above reasoning it follows that such narrow sources should produce anomalous till layers whose thickness is a few cm at best. In reality, these thin ribbons may be spread and diffused into the till by mixing during transport and deposition, which makes it easier to hit the anomaly. Thin ribbon-like anomalies could be lost when applying the typical sample height of 20 cm, if the till section is covered by only a few samples that are placed

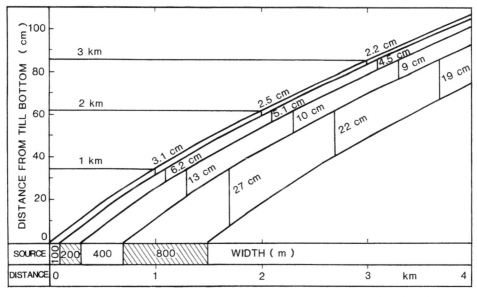

Figure 4. Relationship between the source width and the height of the till layer derived from the source calculated by equation (6) with parameters (P).

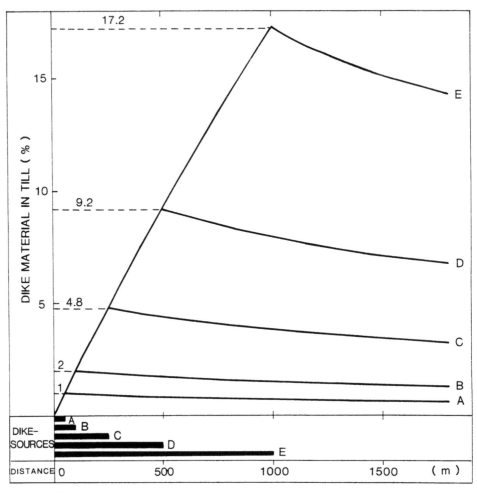

Figure 5. Vertical averages of dike material in a 2-m-thick till layer calculated from equation (8) at parameter values (P). Profiles A-E refer to dike widths of 50-1000 m.

at random. The long mean samples through till sections used in exploration guarantee that the anomaly will be hit. At the same time, however, the proportion of the anomalous layer is reduced in the sample, and the abundances may fall below the anomaly threshold. Longer samples in the vertical direction increase the probability of hitting the anomaly, but reduce the intensity of the anomaly.

In a till sample 20 cm long, a 2-cm-thick till layer produced by a 100-m-wide dike forms only 10% of the sample. The proportion decreases to 1%, when the whole 2 meters till cover is averaged in composition. Figure 5 shows how the profiles of vertical averages change their form, when the width of dike source increases from 50 m to 1 km. It shows that the maximum abundance of dike material in till is always observed at the distal contact of the dike, and rises only to 17%, as the width of the source zone increases to 1 km. The form of till compositional profiles is also strongly dependent on the depth

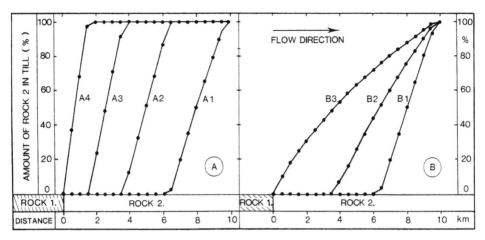

Figure 6. The influence of (A) sampling depth and (B) sample length on a transport profile running over a lithologic contact, calculated from equation (6) with parameters (P) and assuming a till layer 2 m thick. A) The depth interval of samples measured from the layer surface is 0-0.5 m in profile A1, 0.5-1 m in profile A2, 1-1.5 m in profile A3 and 1.5-2 m in profile A4. B) The sample length measured from the layer surface is 0.5 m in profile B1, 1 m in profile B2 and 2 m in profile B3.

and the length of samples. If the depths and lengths are not kept constant along investigation lines, the resulting profiles are difficult to interpret because the abundance variation may be largely due to sampling instead of real transport.

One of the key properties of a model is that it facilitates the planning of strategic sampling. The effect of sampling is illustrated in Fig. 6 where a contact between two rock types has been considered on the basis of transport equation (6). The till cover is assumed to be 2 m thick, and the parameter set (P) is used in the calculations. Shown in Fig. 6A is the influence of the sampling depth on the investigation lines running over the source contact, when a sample, 0.5 m long, is taken from four different depths. As shown by the profiles, the abundance of rock 2 starts to increase in till closer to the contact when the sample is taken deeper (cf. *statement* 4). Figure 6B illustrates the effect of sample length on the abundance profile. The samples taken are 0.5, 1 and 2 m long, measured from the surface of till bed. The abundance of rock 2 starts to rise closer to the contact and rises more gradually, when the sample length is increased. It is obvious from the figures that information on the length and the location of till samples is essential to the interpretation of any till compositional profile.

MODEL APPLICATIONS

The previous theoretic predictions can be applied for geochemical sampling design and for the interpretation of glacigenic anomaly patterns. When the geochemical till mapping aims at the discovery of mineralized bedrock sources the task can be divided into two. First, anomalies must be found in till and, second, the location and size of the cause of the anomaly must be established in the bedrock. The cost-effective realization of the first phase requires planning of sampling, which can be made by using the transport model. Apart from the model and its parameters, the planning needs information about the average thickness of the till cover in the study area. Further, the targets and the anomaly threshold must be specified, and the accuracy of analytical facilities must be known. In order to find the targets with minimum costs and effort, the above information can be used to optimize the sampling grid on lines running in the glacial flow direction.

During glacial transport, material is also spread laterally, as modelled by Häkli and Kerola (1966). The lateral dispersal of material is an essential diluting factor for small point-like (ore) sources. In the case of a 2-dimensional dike or contact, the significance of lateral dispersal is small, unless we are dealing with complex transport characterized by varying transport directions. If the lateral dispersal is small, the sampling line spacing is directly determined by the size of the targets. In the following, we shall examine sampling on lines parallel to glacial transport. The anomaly threshold for the till samples is set at 1%, which implies that samples must contain more than 1% targtet material (or target concentration) before the indication is considered statistically significant.

Let the till layer be 2 meters thick. The targets are defined as dikes, 10 and 100 m in width, trending perpendicular to the direction of glacial flow. The properties of the ice sheet and bedrock are specified by the parameter set (P), implying that the 2-m till cover is renewed over a transport distance of about 10 km. Figure 4 shows that in a till section the material derived from a 100 meters dike is equivalent to a layer about 2 cm thick. Correspondingly, from equation (8) we can calculate that a 10 meters dike produces a till layer 2 mm thick. For the dike material to constitute at least 1% of the till samples, the lengths of the samples cannot exceed $100 \times 2$ mm = 20 cm and $100 \times 2$ cm = 2 m, when we are searching for dikes 10 and 100 m wide, respectively.

The maximum sample lengths are also the optimal lengths when we attempt to localize the dike anomalies with a minimum number of samples. Since the renewal distance of a 2 meters till layer is 10 km, the anomalous dike material in a till can, in principle, be located anywhere within the 10 km stretch. Therefore, when till samples 2 m high are taken along the direction of glacial transport at intervals of 10 km, in theory all the anomalies caused by dikes exceeding 100 m in width are found. In practice it may be difficult to take samples 2 m long. If the sample length is reduced to 1 m, then sample spacing must be changed.

Figure 3 shows that in the 1-m-thick basal part of the till layer the dike anomaly extends for more than 3 kilometers and in the 1-m-thick surficial part of the layer, for about 6 km. Consequently, if anomalies caused by dikes over 100 m wide are to be detected in till, samples 1 m long must be taken systematically either from the till surface at 6 km intervals or from the base of till at 3 km intervals. When searching for dikes 10 m wide the length of the optimal sample was 20 cm. Likewise, from Figure 3 we can deduce that all the anomalies caused by dikes more than 10 m wide will be discovered, in principle, when 20-cm-long samples are taken systematically either from the base of the till at 500 m intervals or from the surface of the till at 1400 m intervals.

Narrow dikes (10–100 m) deliver fairly small amounts of material to till sections, which, when expressed as compact layers, are very thin (2–20 mm). In reality the material is intermixed and a weak anomaly becomes readily discontinuous so that the anomaly is easily missed during systematic sampling. To reduce this possibility it is advisable to take subsamples on both sides of the sampling point proper at, say, 50 m from the point. Since it often costs more to reach the sampling site than to actually take the samples, subsamples are cost-effective. The optimal number of subsamples can be estimated, provided the random variation of till composition and the sampling costs are known.

The above considerations cover the first stage of the geochemical sampling strategy, i.e. the search for and discovery of anomalies in till. In the second stage the detected anomalies are used to predict the location, size and possibly the composition of the source. For that, the vertical location of the anomaly in the till layer must be estimated more accurately with additional shorter samples. The relative amount of anomalous material may increase in the shorter samples, and the maximum values of the anomaly could possibly be used to predict the source composition. The vertical location of the anomaly allows the prediction of source width and position with the aid of suitable diagrams (cf. Fig. 4). The final identification of source formations can be facilitated by comparing the physical properties of till material and provenance bedrock (see Puranen, 1988).

BOULDER TRANSPORT

Tills contain on the average less than 10 wt % stone and boulder fraction with a clast size exceeding 2 cm in diameter (Soveri, 1964). The number of boulders (diameter over 20 cm) is thus low, which makes it difficult to establish their distribution within basal tills. Some of the boulders were destroyed through comminution during glacial transport. As a result, the number of far-travelled boulders is reduced, seemingly shortening the estimates of transport distance. If a boulder has been transported together with basal till material, the model can be used to predict its transport distance. Since the prediction is based on the vertical location of the boulder, the distance will be overestimated if the frost has heaved the boulder towards the till surface. Because of their large size the boulders are prone to the action of frost. However, the boulders have the indisputable advantage over finer till fractions that they are always undiluted samples of the source bedrock even after long transport.

The glacier has usually plucked the boulders from fracture zones or from the lee sides of bedrock hills. In the initial stages of glaciation the small knobs of bedrock could feed many boulders into basal transport. As the bedrock was gradually smoothed by glacial erosion, the abrasion became the dominant mode of erosion instead of plucking, and mainly fine-grained material was taken into transport. The basal entrainment of boulders may have almost ceased at the closing stage of glaciation as is often shown by a gap between the source formation and the proximal boulder of boulder trains. However, the path to the source can be found also in this gap, when finer till fractions are examined. Some boulders have been entrained into transport from the sides of higher hills which rose up sharply into the ice sheet. These boulders were transported clearly above the basal debris and were deposited as part of the ablation till when the ice eventually melted.

Figure 7 shows schematically the entrainment and transport of ablation material from a hill source. In the figure it is assumed that the erosion of material is fairly uniform and that the boulders are fed into the transport as continuous trains. The boulder trains at different levels in the ice illustrate the englacial distribution of material at the end of transport. The far-travelled boulders on the right hand side in the trains were plucked from the hill at the initial stage of glacial erosion. Because the creep velocity within the glacier increases

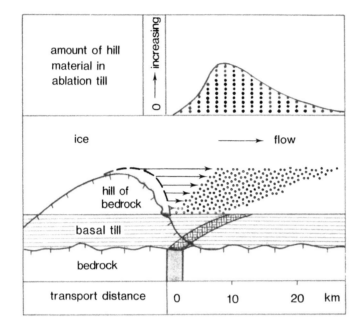

Figure 7. Schematic presentation of transport of ablation till material from a hill source (cf. text).

upwards, the uppermost boulder trains extend the farthest to the right. After the glacier has eroded away the top of the hill, no more material is entrained from there at the final stage of erosion. Therefore, the uppermost boulder train is the first to come to an end on the left. Downwards the following trains extend increasingly close to the source hill, because rock material can be eroded from the lower slopes of the hill until the end of glaciation.

By summing the boulder trains in Fig. 7 we get a typical transport profile, in which the abundance of hill material in ablation till reaches the maximum close to the source and declines gently with increasing distance from the source. The declining part of the abundance curve may be steepened due to lateral dispersion, because hills behave as point-like sources. The shapes of abundance curves can also be affected by comminution, if the boulders are of some weak rock type. If again the boulders are resistant and entrained into rapidly flowing ice layers from high hills, then exceptionally long boulder trains are expected. Discontinuous boulder trains result if boulders are plucked from the source sporadically. Short boulder trains may be due to rapid breaking of boulders, or then boulders may have been fed into glacier only for a short time. The lateral migration due to several hills combined with variations in transport conditions may result in boulder trains that are very difficult to interpret.

It is evident from the above that the source location of boulders can be estimated only within a fairly large distance range. Rough estimates can, however, be made on the basis of the numerous observations compiled by Salonen (1986). Thus, about 50% of the boulders in a typical till cover derive from a distance of less than 3 km, 30% have travelled 3-6 km and 20% more than 6 km. Within these distances the source prediction can be made more accurate in various ways. By determining the physical properties (density, magnetic properties, resistivity, polarizability) of the boulders, geophysical maps can be used to locate the source formation. Topographically the probable source of surface boulders is found on the distal sides of bedrock hills. The boulders in basal till probably derive from smaller rock knobs or fracture zones in bedrock, which can sometimes be observed under the till cover with the aid of impulse radar (Ulriksen, 1982).

DISCUSSION

Except in marginal areas of the ice sheets, the ice flow tends to keep the eroded material at the glacier base. This is especially true in the central areas of the ice sheet, where Finland was located during most of the last glaciation. Therefore, the basal debris is lifted within the ice only by the action of the newly eroded material pushing the previously loosened grains upwards. The lifting velocity depends on the erosion rate of substratum, which is the first component of the transport model for basal tills. In order to erode the substratum, the ice sheet must move by basal sliding. The sliding is the second component of the model, which, apart from eroding, transports material horizontally. The third component is the creep of ice, which also moves the material horizontally. The model does not describe mixing of material during erosion or transport, and deposition of till is simply modelled as the cessation of basal transportation.

The relief of the substratum disturbs the modelling by mixing the till material, and it also may hamper the practical use of transport model. The thickness of the till cover may vary by several meters within a distance of a few hundred meters. As a result highly diverse estimates may be obtained for the distance of a till layer from the bedrock surface, even at adjacent points. The use of these data in the transport model yields contradictory predictions regarding the source location. Therefore, before the model is applied, the observations should be smoothed by calculating within each model increment the mean thickness of till cover and the mean depth of the anomaly. When the sample length, number of samples and the distance increment are appropriate, mean profiles applicable to the model interpretations can be computed. However, the optimal sizes of the samples and distance increments are not yet known, and presumably they vary from one area to the next.

The transport distance of till material depends on the glacier velocities and time. At the glacier base the motion is slow, but upwards in the glacier the creep velocity increases. Therefore, material at a higher elevation in the glacier may travel much farther than the basal debris in the same amount of time. In the central parts of ice sheets the substratum material can enter higher levels of the ice only from topographic elevations. The amount of material being fed from hills into the ice above the basal debris varies, and depends on the relief in the area (Shilts, 1976). The erosion resistance of the glacier bed may also differ in adjacent areas, whereas the time of glacial activities and the glacier velocities depend more systematically on the location of the area in terms of the whole continental ice sheet.

In the areas of the ice-divide in northern Finland the till transport distances are short. The fine till material is often found only a few tens or hundreds of meters from its provenance, while the coarse till fractions, such as stones, have been transported for several kilometers (Lehmuspelto, 1987). This is not easy to explain if we assume that the fines were produced during the transport by comminution from the coarser material.

However, the situation can be understood if the fines belong to the basal till and the coarser material to the ablation till. Stones and boulders are particularly abundant on the topographic elevations of the Lappish bedrock, which was fractured and weathered preglacially. Stones from the hills were pressed into the ice at higher levels, where they could travel farther, depositing eventually as a part of ablation till. In contrast, the finer material was abraded at the glacier base, where the motion was slow and the transport distance short.

The short transport distances in Lapland can be attributed to the low velocities of the glacier at the ice divide. Particularly the basal sliding of the ice sheet was slow, as shown also by the very minor erosion in Lapland. Assuming 20000-30000 years for the duration of the last glaciation in Lapland, we can estimate the glacier velocities there. At transport distances of 250 m (fines) and 2500 m (coarse material) and a duration of glacial activity of 25000 years, we get 0.01 m/a for the velocity of fines, and 0.1 m/a for the coarse material. These velocities are on the low side even for cold-based ice sheets, but the velocity estimates get higher if the glacier was active in Lapland for only part of the glaciation time.

The ablation material in the surface part of till cover always includes some stones and boulders that have travelled farther than the other material: tens or even more than a hundred kilometers (Salonen, 1986). If the rock material in the glacier has reached elevations $z = 3$ or 10 m, the parameter values (P) give $U = kz = 3$ or 10 m/a, respectively, for the creep velocity. At these velocities englacial material had time to travel 30–200 km in the period of 10 000-20 000 years, during which the continental ice sheet probably covered the southern part of Finland. Therefore, the long-distance transport of surface boulders can be explained by the rock material entrained into the ice from the summits of bedrock hills over 3 m high. Rock material from the rapakivi formations in southernmost Finland travelled within the continental ice sheet to the countries south of the Baltic Sea (Hausen, 1912) and there all the way to Holland (Schuddebeurs, 1981). This implies a transport distance of more than 1000 km or multistage transport as proposed by Gillberg (1977).

It can be assumed that the ice sheet advanced from the Scandinavian mountains to southern Finland in 5000 years. It is known that the ice retreated from central Europe back to the Scandinavian mountains in about 10 000 years. If the total duration of the glaciation was 20 000-30 000 years, the glacial flow from southern Finland to central Europe might have lasted for about 5000-15 000 years. during that time rapakivi boulders could have moved 1000 km at a transport velocity of 60–200 m/a. Such velocities are common in present-day glaciers (Paterson, 1981). At the parameter values (P) of the transport model we get $V = B + kz = 60–200$ m/a for the velocity, if the material is raised to a height of 60–200 m above the glacier bed. When the Baltic Sea was covered with thick ice and frozen almost to the bottom, the rapakivi formations of Åland rose from the sea bottom into the ice as hills more than 100 m high in places. From these summits rapakivi material was able to enter glacial transport at elevations exceeding 60 m, at which height the material could reach central Europe during one glaciation.

The presented model offers a conceptual framework for the average movement of basal till material during glacial transport. The model can be used in predicting the source location of different parts of till layers, and also in estimating the amount of till material derived from source areas of varying widths. The closest application will then be in till geochemistry when designing sampling and interpreting glacigenic anomalies. The model considerations emphasize the significance of proper sampling, as the dispersal patterns in till are strongly affected by the length and depth of till samples. Future testing and refinement of the model can be made when new observational data become available. In collecting new data, the specifications of the model should be taken into account.

REFERENCES

Alley, R.B., D.D.Blankenship, C.R.Bentley and S.T.Rooney, 1986. Deformation of till beneath ice stream B, West Antarctica. Nature 322 (3): 57-59.
Bouchard, M.A., B.Cadieux and F.Goutier, 1984. L'origine et les caracteristiques des lithofacies du till dans le secteur nord du Lac Albanel, Québec: une étude de la dispersion glaciaire clastique. In Guha, J. and E.H. Chown (ed.): Chibougamau–Stratigraphy and Mineralization. Canadian Institute of Mining and Metallurgy, Special Volume 34: 244-260.
Boulton, G.S., 1974. Processes and patterns of glacial erosion. In Coates, D.R.(ed.): Glacial Geomorphology. State University of New York, New York: 41-87.
Boulton, G.S., 1979. Processes of glacial erosion on different substrata. Journal of Glaciology 23 (89): 15-37.
Boulton, G.S., 1984. Development of a theoretical model of sediment dispersal by ice sheets. Prospecting in Areas of Glaciated Terrain 1984. The Institution of Mining and Metallurgy, London: 213-223.
Drake, L.D., 1983. Ore plumes in till. Journal of Geology 91: 707-713.
Drewry, D., 1986. Glacial Geologic Processes. Edward Arnold, London. 276 pp.
Echelmeyer, K. and W.Zhongxiang, 1987. Direct observation of basal sliding and deformation of basal drift at subfreezing temperatures. Journal of Glaciology 33 (113): 83-98.
Gillberg, G.,1965. Till distribution and ice movements on the northern slopes of the South Swedish Highlands. Geologiska Föreningens i Stockholm Förhandlingar 86: 433-484.
Gillberg, G., 1977. Redeposition: a process in till formation. Geologiska Föreningens i Stockholm Förhandlingar 99: 246-253.
Häkli, T.A. and P. Kerola, 1966. A computer program for boulder train analysis. Bulletin of The Geological Society of Finland 38: 219-235.
Hallet, B., 1981. General discussion in the Symposium on Processes of Glacier Erosion and Sedimentation. Geilo, Norway 1980. Annals of Glaciology 2: 187-192.
Halonen, O., 1967. Prospecting for asbestos. In Kvalheim. A. (ed.): Geochemical Prospecting in Fennoscandia. Interscience, New York: 171-180.
Hausen, H., 1912. Studier öfver de sydfinska ledblockens spridning i Ryssland, jämte en öfversikt af is-recessionens förlopp i Ostbaltikum. Deutsches Referat: Studien über die Ausbreitung der südfinnischen Leitblöcke in Russland, nebst einer Übersicht der letzten Eisrezession im Ostbaltikum. Fennia 32 (3), 32 pp.
Hellaakoski, A., 1930. On the transportation of materials in the esker of Laitila. Fennia 52 (7), 41 pp.
Hirvas, H., A.Alfthan, E.Pulkkinen, R.Puranen and R.Tynni, 1977. Raportti malminetsintää palvelevasta maaperätutkimuksesta Pohjois-Suomessa vuosina 1972-1976. Summary: A report on glacial drift investigations for ore prospecting purposes in northern Finland 1972-1976. Geological Survey of Finland, Report of Investigation 19, 54 pp.
Hirvas, H. and K.Nenonen, 1987. The till stratigraphy of Finland. In Kujansuu, R. and M. Saarnisto (ed.): INQUA Till Symposium, Finland 1985. Geological Survey of Finland, Special Paper 3: 49-63.
Krumbein, W.C., 1937. Sediments and exponential curves. Journal of Geology 45 (6): 577-601.
Lehmuspelto, P., 1987. Some case histories of the till transport distances recognized in geochemical studies in northern Finland. In Kujansuu, R. and M. Saarnisto (ed.): INQUA Till Symposium, Finland 1985. Geological Survey of Finland, Special Paper 3: 163-168.

Miller, J.K., 1984. Model for clastic indicator trains in till. Prospecting in Areas of Glaciated Terrain 1984. The Institution of Mining and Metallurgy, London: 69-77.

Nurmi, A., 1976. Geochemistry of the till blanket at the Talluskanava Ni-Cu ore deposit, Tervo, Central Finland. Geological Survey of Finland, Report of Investigation 15, 84 pp.

Paterson, W.S.B., 1981. The Physics of Glaciers. Pergamon Press, Oxford. 380 pp.

Peltoniemi, H., 1985. Till lithology and glacial transport in Kuhmo, eastern Finland. Boreas 14: 67-74.

Perttunen, M., 1977. The lithologic relation between till and bedrock in the region of Hämeenlinna, southern Finland. Geological Survey of Finland, Bulletin 291, 68 pp.

Pulkkinen, E., R.Puranen and P.Lehmuspelto, 1980. Interpretation of geochemical anomalies in glacial drift of Finnish Lapland with the aid of magnetic susceptibility data. Geological Survey of Finland, Report of Investigation 47, 39 pp.

Puranen, R. 1988. Modelling of glacial transport of basal tills in Finland. Geological Survey of Finland, Report of Investigation 81, 36 pp.

Salminen, R., 1980. On the geochemistry of copper in the Quaternary deposits in the Kiihtelysvaara area, North Karelia, Finland. Geological Survey of Finland, Bulletin 309, 48 pp.

Salminen, R. and A.Hartikainen, 1985. Glacial transport of till and its influence on interpretation of geochemical results in North Karelia, Finland. Geological Survey of Finland, Bulletin 335, 48 pp.

Salonen, V.-P., 1986. Glacial transport distance distributions of surface boulders in Finland. Geological Survey of Finland, Bulletin 338, 57 pp.

Sauramo, M., 1924. Tracing of glacial boulders and its application in prospecting. Bulletin de la Commission géologique de Finlande 67, 37 pp.

Schuddebeurs, A.P., 1981. Results of counts of Fennoscandinavian erratics in the Netherlands. Mededelingen Rijks Geologische Dienst 34 (3): 10-14.

Shilts, W.W., 1976. Glacial till and mineral exploration. In Legget, R.F. (ed.): Glacial Till. Royal Society of Canada, Special Publication 12: 205-224.

Soveri, U., 1964. Maalajit ja niiden käyttö. In Rankama, K. (ed.): Suomen geologia. Kirjayhtymä, Helsinki: 333–376.

Ulriksen, P., 1982. Application of impulse radar to civil engineering. Lund University, Department of Engineering Geology, Doctoral Dissertation. 179 pp.

Virkkala, K., 1971. On the lithology and provenance of the till of a gabbro area in Finland. VIII International Congress INQUA, Paris 1969. Etudes sur le Quaternaire dans le Monde: 711-714.

# CHAPTER 3

# Formation, deposition, and identification of subglacial and supraglacial tills

ALEKSIS DREIMANIS

*Department of Geology, University of Western Ontario, London, Canada*

## INTRODUCTION

Most ore deposits are found in bedrock. In areas that have been glaciated during the Quaternary, bedrock is commonly concealed under so called 'overburden' that consists mainly of glacigenic (glacially-derived) sediment. Quaternary glacigenic sediment is genetically completely different from the underlying bedrock, but lithologically they usually have some similarities. Still, they are separated by an unconformity, and overburden is often considered to be a hindrance to bedrock mapping and prospecting for ore deposits.

Glaciers, the producers of glacigenic sediment, derive their debris from bedrock or by recycling older sediment that, in turn, has also been derived from bedrock. If the processes of glacial transport and deposition, and the preceding transport of the glacially recycled debris can be correctly deciphered, then overburden no longer is a hindrance, but it becomes a source of information. This has been realized in Finland as early as the first half of the 18th century by Tilas (1740). Ore boulders found in overburden were used by him for tracing them to their source outcrops. If investigated in detail glacial sediment becomes a valuable indicator in the search for ore deposits and in bedrock mapping.

When using glacigenic deposits as a source of information in prospecting the emphasis is on their lithology and geochemistry. From the prospector's point of view the derivation of ore-bearing glacial debris and their transport, particularly the direction and distance of transport, are the main genetic parameters of interest. The usual source of information is the glacial deposit and, if it is a surface deposit, then also its landform. However, their specific mode of formation and deposition may include some information on their predepositional transport and derivation.

For instance, if ore fragments are found in a fluted groundmoraine terrain, in a region that has been glaciated during the Pleistocene by a continental ice sheet, then their subglacial or englacial transport is suggested by the landform and its main glacial deposit, basal till. (To shorten the discussion of this example, let us assume that the basal till and the fluted groundmoraine have been correctly identified by multiple descriptive criteria.) If ore fragments are found only on the till surface, but none in the till itself, then a long-distance englacial transport is most likely. However, if they are also present in the basal till, then the distance of glacial transport may be relatively short, unless the ore fragments have been recycled from sediments older than the till sampled.

Without going into methodological details, the following criteria for further investigation shall be mentioned, in a situation where only one basal till sheet in contact with

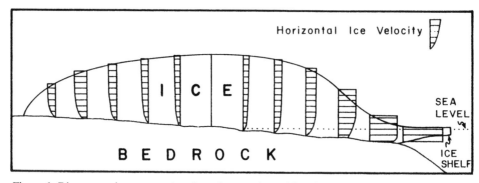

Figure 1. Diagrammatic cross section through a continental ice sheet with terrestrial (left side) and marine (right side) portions. The vertical scale is strongly exaggerated. After Hughes (1981: Fig. 5-2).

bedrock has been found, and where ore fragments and/or geochemical anomalies have been determined in the till. The direction of the latest local glacial movement is deciphered from the orientation of the surface flutings, and the sense of the movement – from the stoss-and-lee relationships of erosional features on several glacially abraded boulders. Measurements of the orientation of clasts in the till (the so called 'till fabric') and of sets of parallel striae and other erosional marks on the underlying bedrock, also on clast pavements if they are present, will supply additional information on the local direction of glacial movement during the deposition of till. These measurements should be done at several places and at several depths, considering the multiple processes involved in the deposition of basal till (Appendix A-2) and a possibility that the direction of glacial movement had changed during the deposition of till.

Quantitative lithologic and geochemical investigation of till, and a comparison with the bedrock lithology in the upglacier area, will assist in the determination of the regional ice flow direction and in finding the possible provenance area of the ore fragments. If the process of deposition of each subunit of the basal till can be deciphered, it will indicate the relative distance of glacial transport. For instance, deformation till or the basal part of a lodgement till, if rich in local bedrock material, will suggest a small transport distance, but if the ore fragments are present in melt-out till overlying lodgement till, they may have travelled a variable distance (for more details see below).

As many criteria as possible shall be used for crosschecking the genetic, directional, and distance interpretations, and the entire investigation should proceed like three-dimensional regional mapping. Factual descriptive data must be collected first, and their genetic and other interpretations follow with repeated crosschecking, as the mapping area expands.

From the prospector's point of view glacier ice is the most important geologic agent in a glacigenic environment and till is the most informative sediment in which to search for ore fragments and geochemical anomalies. Good understanding of glacial geologic processes, particularly those which are related to the formation and deposition of till, are needed. For general information following publications are useful: Ashley, Shaw and Smith (1985), Sugden and John (1976), Brodzikowski and van Loon (1987); a more advanced glaciologic information is given in Drewry (1986).

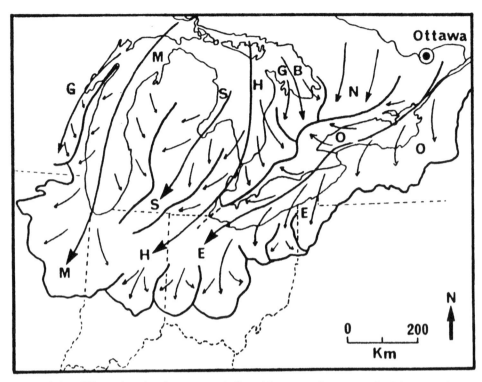

Figure 2. Late Wisconsinan ice-flow patterns indicated by arrows; ice streams, ice lobes and sublobes in the Great lakes region, North America. Abbreviations from left to right: G – Green Bay lobe, M – Lake Michigan lobe, S – Saginaw Bay lobe, H – Huron lobe, GB – Georgian Bay lobe, E – Erie lobe, O – Ontario lobe, N – northern ice (after Barnett and Kelly 1987: Fig. 4).

Every prospector should be aware also that glacier ice is not the only geologic agent in the glacigenic environment, and several other geologic agents may deposit till-like diamictons. Also post-depositional processes may affect the appearance and, particularly, the geochemical composition of glacigenic sediment. All these complications are beyond the scope of this paper and will not be discussed here.

MAIN GLACIER TYPES

Several types of glaciers are distinguished, according to their morphologic form and relation to the topography (Flint 1971: 28-51; Sugden and John 1976: 56-71). The most common varieties are ice sheets and valley glaciers.

An ice sheet (Fig. 1) is independent in its general form from the underlying topography, and the glacial movement is determined mainly by its surface gradients. However, locally the movements are guided by the underlying topography, and internal ice streams develop in an ice sheet. They may emerge at the periphery as glacial lobes and sublobes (Fig. 2) or as outlet glaciers. Large ice sheets usually develop several domes of outflow.

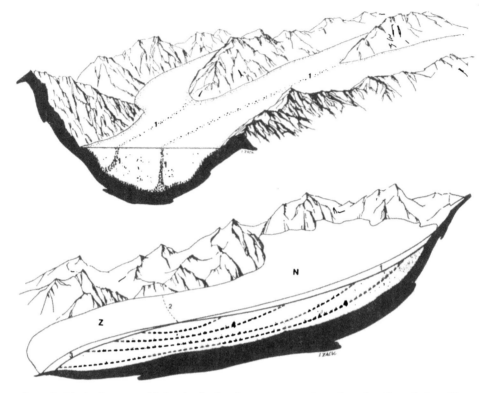

Figure 3. Block diagrams with longitudinal and transverse sections through valley glaciers. Upper diagram: (1) medial moraine. Lower diagram: (N) area of accumulation with (1) snow and firn, and (2) equilibrium line below which is (Z), the area of ablation; (3) represents the mass of ice lost by surface ablation; (4) flowlines in ice. Both block diagrams are from van Husen (1987: Figs. 5 and 8).

Both ice domes and ice streams shift their positions during the life time of an ice sheet, consequently causing complications in indicator tracing.

An ice sheet may occupy any part of a continent (a continental or terrestrial ice sheet) including inland seas, for instance the Baltic Sea during the Pleistocene glaciations, or it may even rest on a continental shelf, then being called a marine ice sheet (Fig. 1).

Valley glaciers (Fig. 3) are constrained by the topography of mountain valleys. If they merge laterally and overwhelm the mountains, leaving only some high mountain peaks as nunataks above the ice, they develop into a mountain ice sheet as in Antarctica.

If glaciers terminate in large water bodies, icebergs break off and spread glacial debris into aquatic sediments. Where glaciers, particularly ice sheets, end is deep water, their termini may float as ice shelves (Fig. 1: right side).

TILL, RELATED TERMS, DEFINITION OF TILL

The Scottish term 'till' is the most popular name for glacial sediment in the English language and it is gradually invading the scientific terminology of other languages (Dreimanis 1989: Appendix C). Still, the French Alpine term 'moraine', in various spellings and combinations with compositional terms, dominate in most European languages, for instance 'Moräne' or 'Moränenmaterial' in German, 'moreeni' or 'moreeniaines' in Finnish. Descriptive terms, still meaning a glacial deposit, are also commonly used; for instance 'boulder clay' in English, 'Geschiebemergel' in German, 'Keilehm' in Dutch terminology. Descriptive terms without glacial connotation, meaning poorly sorted sediment containing a wide range of particle sizes from clasts to clay size, are 'diamicton' (unconsolidated), 'diamict' (general), 'diamictite' (consolidated or cemented), and 'mixtite' (usually consolidated).

Though till is deposited by or from glacier ice and consists of debris that has been transported or at least moved by glacier ice, it is actually a polygenetic sediment. Except for a little studied variety of ablation till, deposited by sublimation, all other tills are formed and deposited by processes that involve melting of glacier ice and production of meltwater. Gravity also participates, as in any other sedimentation process.

Detailed statistical granulometric investigations of tills of various genetic origins formed by various types of glaciers, and consisting of a variety of lithologic materials, demonstrate that all of them are poorly to very poorly sorted, as shown by Braun, German and Mader (1976), Mills (1977), Vorren (1977), van der Meer (1982), and Rappol (1983). Apparently, the subsidiary agents, meltwater and gravity, have not achieved much sorting during the formation of tills. Increase in sorting in till results more from incorporation of water-sorted or wind-sorted sediment.

If the broad genetic definition of till, as proposed by the INQUA Commission on genesis and lithology of Quaternary deposits (Dreimanis 1982) is supplemented by its most typical petrographic characteristic, the diamictic texture, then the following petrogenetic definition may be useful to prospectors:

Till is a diamicton that has been transported and is subsequently deposited by or from glacier ice, with little or no sorting by water. As pointed out by Dreimanis and Lundqvist (1984: 9), this definition demands that the following three restrictive specifications have to be met, in order to call a diamictic sediment a 'till':

(1) till consists of debris that has been transported by glacier;
(2) close spatial relationship exists to a glacier:
– till is deposited by a glacier, or
– it is deposited from a glacier;
(3) sorting by water is absent or minimal during the formation of till.

The field characteristics of till are discussed at length by Dreimanis and Schlüchter (1985: 9-12).

DERIVATION OR ENTRAINMENT OF GLACIAL DEBRIS

Glacial and meltwater erosion at the base of ice occurs in all glaciers, and it is the principal source of glacial debris in continental ice sheets (Fig. 4). Wind contributes some fine extraglacial or exogenous (non-glacially derived) particles, mainly of silt size, to the surface of ice. Locally, around volcanoes, tephra may also land on the surface of glaciers.

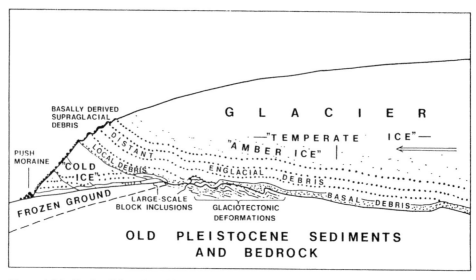

Figure 4. Idealized radial section of ice sheet in its terminal zone. Arrow indicates direction of glacial movement. After Dreimanis (1976: Fig. 2).

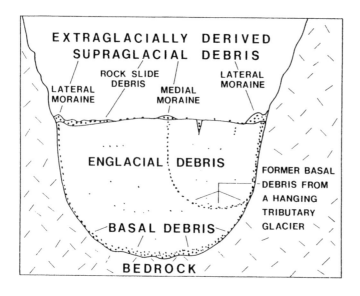

Figure 5. Idealized cross section of a valley glacier in the area of ablation. Glacial debris is shown by heavy dots.

Mountain glaciers receive considerable amounts of exogenous (or extraglacial) debris from the valley walls above the glacier ice surface (Fig. 5) and some of their englacial, glacially derived debris also derive from valley sides.

A distinction of all the above modes of derivation or entrainment of glacial debris is important for prospectors, since each of them means a different probable source for the prospecting object.

GLACIAL TRANSPORT

Since glacial transport is of considerable interest to prospectors, it will be briefly reviewed here (see Figs. 3-7), beginning with the lowermost position of transport.

## Subsole drag

Where a glacier is underlain by water-saturated unconsolidated sediment or soft or incompetent bedrock, for instance soft shale or grussified granite, such material may be transported by subsole drag without direct incorporation into the glacier ice. About a century ago, the Alpine glacial geologists regarded the subsole drag as one of the major processes of glacial transport prior to the deposition of basal till or 'Grundmoräne' (for further references see von Böhm 1901). During the first half of this century subsole drag was disregarded until its reintroduction as a process of the formation of deformation till by Elson (1961). Boulton and Jones (1979) proved it experimentally under Breida-merkurjökul in Iceland, and several recent detailed investigations of till sections indicate that it is a common process, particularly in poorly drained lowland areas (see Grube and Vollmer 1985, Barnett 1987, Boulton 1987, Boulton and Hindemarsh 1987, Dreimanis, Hamilton and Kelly 1987, and a discussion in Dreimanis 1989: 9.2.4).

## Basal transport in ice

Above the glacier sole is the basal zone of traction in the glacier ice (Figs. 4-6). Most of the basally entrained glacial debris is transported here.

The basal zone of transport is present in all glaciers, but its thickness may vary from erratic occurences of scattered particles in some mountain glaciers to several metres. (For further references see Drewry 1987: 97-102.) The debris concentration in the basal zone averages about 25% by volume, but it may range from a fraction of a percent in ice layers containing only dispersed debris, to 90% in distinct debris-rich dark layers (Lawson 1979; Pessl and Fredrick 1981). Most glacial comminution and abrasion takes place in this zone or along the glacier sole.

## Englacial transport

The basal zone of transport in glacier ice is overlain by the debris–poor englacial zone (Figs. 4-6) that contains suspended diffused particles and aggregates, and is commonly called 'amber ice'. In the areas of compressive flow thin debris-rich bands become raised up by the compressive flow from the basal zone. In the areas of strongly crevassed ice, for instance at ice falls in mountain glaciers, supraglacial debris fall into the englacial zone along the crevasses. Since crushing is unlikely and abrasion is minimal in the englacial zone, weak rock fragments and soft sediment lumps may survive a lengthy transport in it (Dreimanis 1976: 21-23).

## Supraglacial transport

Supraglacial transport is most important in mountain glaciers (Fig. 5) where mass movements of extraglacial debris from valley sides are the main contributors of supraglacial exogenous debris (Sharp 1949). They are carried passively, either exposed on the glacier surface or buried in snow, firn, or in the upper brittle part of ice (Eyles 1979). Therefore, these supraglacial debris do not show any glacial abrasion marks. Some fines however, become washed out by percolating water.

On continental ice sheets, the main area where dark glacial debris are noticeable is along the glacier margin (Fig. 4). Here the surface melt-out concentrates englacial debris on its surface. Because of the previous basal and englacial transport the clasts are glacially abraded.

Aeolian debris dropped on any glacier surface are carried passively, without any signs of glacial abrasion.

## *Glacial buldozing by the terminus*

Glacial buldozing along its terminus, forming push moraines (Fig. 4), has been recognized along present-day glaciers, for instance by Rabassa, Rubulis and Suarez (1979), and Evenson and Clinch (1987). It has been considered more as a local redepositional and deforming (Croot 1987), rather than transporting, process.

If ore fragments are found in a push moraine consisting of ice-marginal, either 'glacier sourced' or 'inwash sourced' material (Evenson and Clinch 1987), their transport is related more to the transporting agent of the buldozed material than to the short-distance resedimentation by buldozing.

## *Lithologic differences among glacial transport zones*

After deciphering the genesis of till, its facies associations, and landforms, it is often possible to conclude what type of glacial transport preceded the deposition of till. Therefore, I will review and summarize those lithologic characteristics of the above discussed principal zones of glacial transport, which permit an estimation of relative length of transport, and its complexity or lack thereof.

Local materials dominate the subsole and the basal zone of glacial transport. The upper part of a thick basal zone, particularly in cold-base glaciers, and the entire englacial zone contains more distantly transported material. Clasts transported in all the above three zones usually bear glacial abrasion marks: rounding of edges, striae, etc., however they are most strongly developed in the basal zone of traction.

The supraglacial exogenous clasts may be transported on and within a glacier surface any distance without having been affected by glacial abrasion, or some of them may become rounded by supraglacial meltwater streams in the area of ablation. Most of them remain angular, or even increase their angularity by frost-shattering. If supraglacial debris has been transported in medial moraines it preserves its original lithologic composition of the area of derivation and, therefore, is very useful in prospecting (Stephens et al. 1983). A certain amount of englacial debris of various derivation also becomes admixed to the supraglacial zone of transport in the area of surface ablation, and in areas of compressive flow. However, englacial debris may be distinguished from supraglacial debris by its surface abrasion features.

Recycling of old unconsolidated sediments (Gillberg 1977) may occur in any zone of transport, depending upon their availability for incorporation along the glacier's path. The greatest possibility of encountering them is in the zones of subsole drag and basal transport.

ENVIRONMENT AND PLACE OF DEPOSITION OF TILL

Though till is commonly considered to be a terrestrial or glacioterrestrial sediment (Boulton and Deynoux 1981), it may also be deposited subaquatically, for instance in water-filled cavities underneath a glacier, or along the glacier margin terminating in a lake or sea.

In relation to glacier ice, the position of deposition of till is either subglacial, supraglacial, or ice-marginal. The ice-marginal and supraglacial positions are usually considered together (Fig. 6), since supraglacial deposition most commonly occurs in the ice-marginal area, as a result of ablation of glacier ice.

PROCESS OF DEPOSITION OF TILL

Two principal groups of processes participate in the polygenetic deposition of till: (1) primary, and (2) secondary.

From the glacial sedimentologic point of view, the primary glacial processes are those which are genetically directly related to the glacier; that it, where the glacier ice has been the principal depositional agent. The following three primary depositional processes are currently recognized: (1) melting out and sublimation, (2) lodgement, and (3) stoppage of subsole drag. They often interact among themselves and also with secondary processes.

Secondary processes, such as mass movements and water movements, cause resedimentation of glacially- and non-glacially-derived materials in the glacial environment. If they operate completely independently of the glacial primary processes, they are usually not regarded as till-forming processes. However, if they participate in the primary deposition of till, or if they follow the primary deposition in a continuum without significantly altering the original composition of the primary till, then they are still considered as till-forming processes by most glacial geologists, but not by all. This currently controversial matter will be further discussed below.

*Melting out or melt-out, and sublimation*

'Melt-out is the slow release of debris from glacier ice that is not sliding or deforming internally...Melt-out occurs in some places from ice bodies that are wholly stagnant. Elsewhere it occurs from stagnant, basal ice that is separated by a sliding plane from overlying, active ice' (Shaw 1985: 38). Melt-out is a primary passive ablation process in the glacial environment and it takes place both at the surface and the base of a glacier.

In extremely cold and dry climate melt-out may be replaced by sublimation (Shaw 1977) at the surface of glacier ice and in some ice caves. Since sublimation may combine with supraglacial melt-out, its product, sublimation till, is included with supraglacial melt-out till in Fig. 6.

*Lodgement*

Lodgement is deposition of till from the sliding base of a moving glacier by pressure melting and/or other mechanical processes. Lodgement is accomplished (1) particle by particle released from the glacier sole, or (2) by shearing and plastering thin sheets of

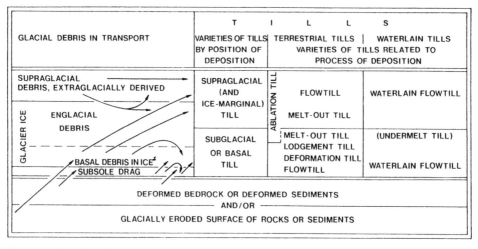

Figure 6. Simplified depositional genetic classification of tills, and their relationship to the transport and derivation of glacial debris as indicated by arrows. Sublimation till is included with the supraglacial melt-out till. For the differentiation of exogenous from basally derived supraglacial till see Figure 9. The terms in parentheses ( ) are used less frequently.

basal melt-out debris, or (3) by lodging entire sheets of debris-rich basal ice that later melts out.

Lodgement has been considered by many glacial geologists to be the most important primary glacial depositional process under a dynamically active wet-base glaciers. Because of the difficulties in observing the slow and hidden process of lodgement under moving glaciers, the variety of mechanisms of lodgement are still being debated (Dreimanis 1989: 9.2.1).

*Stoppage of subsole drag and squeenze-flow*

Strongly deformed local weak bedrock or unconsolidated substratal material that has been detached from its original source by subsole drag, and transported for some distance, will eventually be deposited. This will occur when the force imposed by the overriding glacier ice becomes insufficient to maintain the forward motion of the material dragged. This process appears to be similar to the lodgement process discussed above. The main difference, theoretically, is that lodgement deposits basal debris shortly after its release from the glacier sole, while subsole drag and its stoppage may affect any subsole material, glacial or non-glacial. The stoppage of subsole drag and shearing may actually be part of the lodgement process, as pointed out by Sugden and John (1976: 217) and elaborated in many more recent papers (for further references see Dreimanis 1989: 9.2.4).

Water-saturated subsole material may also be squeezed by the load of the overlying glacier ice into any open or less confined spaces: into cavities, crevasses, etc. This is a primary process of resedimentation, since it is produced by glacier ice. It results in the formation of 'squeeze-flowtill' that is similar in its properties to gravity flowtill, a product of secondary resedimentation (for references see Dreimanis 1989: 9.2.4).

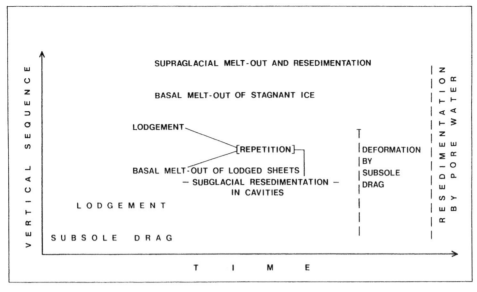

Figure 7. Multilevel deposition of till in a continental ice sheet close to its margin. After Dreimanis (1989: Table 13).

## Resedimentation by gravity

Resedimentation by gravity as a secondary till-forming and depositing process may occur either (1) locally and selectively inside a primary till, or (2) as a continuous complete or partial resedimentation of a primary till or glacial debris just being released from the ice, at the same glacier where the primary process had started the formation of till.

An example of localized and selective participation of secondary processes inside a primary till is the downward resedimentation of some fine particles in voids during melt-out (discussed in Dreimanis 1983).

A continuous complete or partial resedimentation occurs as downslope flow, slumps and/or sliding, involving debris just released from glacier ice, or even an already deposited primary till that is not stable enough to remain in its original position. If the flowage is very slow, a gradual transition exists from the primary till and its resedimented portion (Lawson 1979: 104; Dreimanis 1987).

Differences in opinions exist as to whether to consider such resedimented parts of a till unit a true till, or to call them a non-till. In the first case the name 'flowtill', in a broad meaning, may be applied to the resulting sediment (Dreimanis 1989 and references therein). However, if the second option is preferred it may be called a 'sediment flow deposit' (Lawson 1979 and 1989), a 'glacigenic massflow deposit' (Shaw 1985: 47), or a 'flowed till' (H.J. Stephan's proposal, personal communication, 12.10.79, distributed to INQUA Till Work Group 1980). If the resedimented portion of glacial debris or till is distinguishable as a mappable unit, its separation from primary till as a genetically and sedimentologically different non-till subunit is feasible. However, very often the resedimented diamiction interdigitates with the related primary diamicton so closely, that their separation is difficult. Since lithologically and geochemically both subunits would be similar anyway, it would be simpler, from the prospector's point of view, to consider both the primary and the secondary portions as parts of the same till unit if they derived from the same glacier, as has been done, for instance, by Barnett (1987), Dreimanis, Hamilton and Kelly (1987).

## SIMULTANEOUS DEPOSITION AT SEVERAL LEVELS

Glacial deposition may be in progress simultaneously at several levels (Fig. 7): beneath the glacier sole, within the glacier, and on its surface. The most commonly recognized contemporaneous pair is subglacial and supraglacial sedimentation in an ice-marginal area that results in the deposition of subglacial or basal till overlain by supraglacial till, usually called ablation till by prospectors (Figs. 10 to 13). Each of these two till units, in turn, may consist of several simultaneously- or penecontemporaneously-deposited, genetically different, subunits that may be distinguished as different facies by detailed investigation. The following references may be mentioned, each of them showing a different set of combinations, or some variability in a similar set-up: Eyles, Sladen and Gilroy (1982); Muller (1982); Ruszczyńska-Szenajch (1982); Åmark (1985); Dreimanis, Hamilton and Kelley (1987); Hansel, Johnson and Socha (1987); Shaw (1987).

The following penecontemporaneous processes participate in the formation and deposition of a subglacial till unit:

1. Lodgement by various mechanisms;
2. Basal melt-out;
3. Gravity flow and squeeze flow in subglacial cavities;
4. Meltwater sedimentation and free fall from the ice roof in subglacial or englacial cavities forming non-till lenses in till;

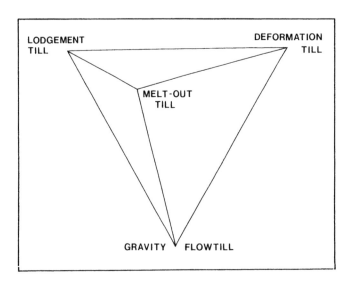

Figure 8. Schematic depositional genetic classification of till, with four principal endmembers at the apices of a tetrahedron. After Dreimanis (1989: Fig. 1).

5. Deformation of any of the sediments produced by (1-4), by glacial drag resulting either in glacitectonic deformation structures or in the formation of deformation till;

6. Resedimentation of some fine particles by pore water in intersticies, particularly in coarse-textured tills with large voids;

7. Intermittent erosion locally removing some of the already deposited till and abrading the surface of clasts along the intratill erosional surfaces.

The following processes participate in supraglacial deposition of till and related non-glacial sediments:

1. Surface melt-out and sublimation;

2. Gravity flow, slumping, and free fall along steep ice slopes, resulting in resedimentation of various magnitude;

3. Small-scale resedimentation by pore water in voids;

4. Localized meltwater sedimentation producing glaciofluvial and pond deposits;

5. Aeolean sedimentation (insignificant);

6. Deformation of any of the sediments produced by (1-4), by normal faulting and slumping, due to melting of buried ice underneath;

7. Buldozing of ice-marginal deposits along the glacier front during glacial readvances;

8. Localized erosion, mainly by meltwater.

The interaction of the variety of glacigenic depositional, deformational, and erosional processes, shifting and alternating both in space and time, and their polygenetic character makes it difficult to develop a genetic classification of tills that would apply to all situations.

## GENETIC CLASSIFICATION OF TILL, FROM THE PROSPECTOR'S POINT OF VIEW

Most genetic classifications of till are depositional, and they are usually based upon the processes of deposition, the position of deposition in relation to glacier ice, and the general environment (terrestrial or aquatic) where tills are deposited. The gradual development of till classifications during the last hundred years is discussed with further references in Dreimanis (1989), and a simplified classification is given in Figure 6.

The currently most commonly used terms for the genetic depositional process varieties of till, such as lodgement till, melt-out till, deformation till, and gravity flowtill, are actually extreme end-members in a system resembling a tetrahedron (Fig. 8). Most tills are combinations of them and they may be placed in hundreds of positions within the tetrahedron. Still, the terms of the end members are commonly used in many reports, if the particular process appears to be dominant in the formation of the till unit examined. However, that is commonly an oversimplification or even a hypothetical assumption based upon one of the facies of the till unit studied, or upon only some criteria used for genetic classification.

From the prospector's point of view, the process of deposition is not as important as the position of deposition, in relation to glacier ice, and in relation to large water bodies such as proglacial lake or sea. Also, the predepositional transport and derivation of glacial debris (Fig. 9) is important to prospectors, and recycling of older sediments shall also be considered.

| DERIVATION OF GLACIAL DEBRIS | TRANSPORT OF GLACIAL DEBRIS | GENETIC VARIETY OF TILL |
|---|---|---|
| EXTRAGLACIALLY DERIVED | SUPRAGLACIAL, ENGLACIAL | EXOGENOUS SUPRAGLACIAL TILL |
| BASALLY DERIVED | SUPRAGLACIAL, ENGLACIAL, BASAL | BASALLY DERIVED SUPRAGLACIAL TILL |
| | ENGLACIAL, BASAL, SUBSOLE | SUBGLACIAL OR BASAL TILL |

Figure 9. Relationship of three genetic varieties of till (right side column) that are of special interest to prospectors, to their transport (central column) and derivation (left side column).

Figure 10. Till section in the proglacial area of Solheimajøkull, Iceland. The lower half is a fissile dense subglacial till, probably a lodgement till, with clasts aligned parallel to the glacial movement from the left to the right. The loose textured upper half, with pebble fabric unrelated to the glacial movement, is probably a supraglacial (ice-marginal) flowtill, since a concentration of clasts separates it from the underlying subglacial till; it could be interpreted also as a resedimented dilated surface portion of the subglacial till.

Figure 11. An Illinoian drift section in the Mill Creek valley at Cincinnati, Ohio, U.S.A. with its left side being S.W.: (A) – stratified glaciolacustrine clay and silt, with thin diamicton interbeds; (B) – dense massive subglacial clayey silt till with shear planes rising downglacier (S.W.); (C) – loose coarse textured supraglacial till (surface ablation till), probably an ice-marginal flowtill. Glacial movement was from the right towards the left.

The following two major groups of tills (Figs. 6 and 10-13) are recognized according to their relationship to a glacier:

(1) *Subglacial or basal till* that includes several genetic depositional varieties: lodgement till, deformation till, squeeze flowtill, subglacial gravity flowtill, and subglacial or basal melt-out till (Figs. 10–12),

(2) *Supraglacial or ice-marginal till*, known to most prospectors as *'ablation till'*. Its main depositional varieties are supraglacial melt-out and sublimation till, and gravity flowtill (Figs. 10, 11, 13).

The use of the term 'ablation till' has been discouraged for various reasons (Boulton 1976; Dreimanis 1989). One of them is that 'ablation' means a process, and not a position; another, basal melt-out till is also formed by ablation. It might be even difficult to distinguish basal melt-out till from a supraglacial melt-out till. The term 'ablation till' has been so strongly entrenched among field geologists and prospectors, that it is still the dominant designation for supraglacial till, if it is found to be different than the related basal till. If used as a positional term, it should be specified as 'surface ablation till'.

Another parallel term, 'ice-marginal till' is used besides 'supraglacial till' (Dreimanis 1989), because most flowtill becomes deposited in the ice-marginal area, adjacent to the glacier from which it originates. However, after the ice melted it is commonly hard to tell wether the flowtill was deposited on glacier ice, on a snowbank, or on ground adjoining to ice. Since its formation begins on ice anyway, starting with the supraglacial melt-out of debris or sliding of supraglacial debris, the term 'supraglacial till' may be used in most cases as a positional term.

Figure 12. Basal till from Finnish Lapland. Photo R. Kujansuu 1973.

Figure 13. Surface ablation till from Finnish Lapland. Photo R. Kujansuu 1971.

In the depositional group of supraglacial or surface ablation till two endmembers may be distinguished, when considering the derivation and transport of glacial debris (Fig. 9):

(2a) *exogenous supraglacial till*: its debris have been derived by the activity of nonglacial exogenous or extraglacial agents, mainly by various mass movements from the adjoining valley sides and nunataks (Figs. 3 and 5), and transported passively without the abrasive action of glacier ice, either supraglacially or in the upper part of the ice;

(2b) *basally derived supraglacial till*: its debris were eroded at the base of glacier ice then transported basally and englacially and subjected to glacial abrasion, and finally released supraglacially by the melt-out process (Fig. 4).

Continental ice sheets produce only variety (2b), while mountain glaciers deposit a combination of (2a) and (2b), with the variety (2a) usually dominating. Boulton (1976: 72) named the variety (2a) 'supraglacial morainic till'. It is difficult to translate this term in the languages where till is called moraine, and therefore the Till Work Group of INQUA Commission on Genesis and Lithology of Quaternary Deposits proposed in 1980, following a suggestion of J. Shaw and E. Lagerlund, to call it 'exogenous till' (Dreimanis 1982: 28). Boulton and Deyonoux (1981: 403) proposed another term, 'supra-till', as a replacement of 'supraglacial morainic till'. However, 'supra-till' is not the only supraglacial till, and when classifying supraglacial tills into the above varieties (2a) and (2b) it would be cumbersome to call one of them a 'supra-till' or 'supra-supraglacial till', and the other a 'basally derived supraglacial till'. Therefore I would propose retaining the term 'exogenous (supraglacial) till' for variety (2a). For those who prefer to use the term 'ablation till' instead of 'supraglacial till', the substitutes for (2a) would be 'exogenous ablation till' and for (2b) – 'basally derived ablation till'.

A question arises, as to the application of the above classification (Figs. 6 and 9) to tills deposited in water adjacent to glacier ice. They have been often considered as a separate group called waterlain tills, subaquatic tills, glacioaquatic tills, or aquatills (for reviews see Dreimanis 1979 and 1989: 11.2). Since most of them are deposited by mass movements, particularly gravity flow, Dreimanis (1988: 11.2) proposed that they be included with flowtill in a broad meaning of this term. A specification 'waterlain' or 'subaquatic' could be added to the positional or depositional genetic variety of till, for instance

– 'waterlain ice-marginal till' = 'waterlain ice-marginal flowtill'.

In the case of waterlain subglacial till a question may arise whether a diamicton deposited by free fall from the sole of the glacier in water as a product of basal melting (Gibbard 1980) and named 'undermelt diamicton' by Gravenor, von Brunn and Dreimanis (1984), is a till or a glaciolacustrine sediment. If it is considered a till (Gibbard 1980; Parkin and Hicock 1989), then it belongs to the subglacial or basal till group, by its position of deposition (Fig. 6).

*Criteria for recognition of the genetic varieties of tills*

No single criterion is sufficient to identify any of the genetic varieties of till. Multiple criteria are needed and even they may overlap or vary from one situation to another. Since prospecting for ore deposits usually begins with field work, descriptive field criteria are most useful for the identification and classification of till; for example the criteria discussed in Appendices A-2 and A-3. They deal with the two position-related genetic varieties of till – subglacial and supraglacial tills. The most recently published descriptive criteria for lodgement, melt-out and gravity flowtills are given in Dreimanis (1989 : Appendix C), for deformation till in Elson (1989), sublimation till in Shaw (1989), and protalus till (a variety of supraglacial till) in Warren (1989). For older references see Dreimanis (1989).

ACKNOWLEDGEMENTS

While the main part of this report expresses the author's views on those genetic parameters of tills that appear to me to be important to prospectors, my opinions have been influenced by those members of the Till Work Group of INQUA Comission on Genesis and Lithology of Quaternary Deposits, who have actively participated in the discussions on the descriptive properties of subglacial or basal till versus supraglacial and/or ablation till during the period of 1978-1986. They should be considered as co-authors of the Appendices A-2 and A-3, and therefore their names are listed below:

R. Aario (Finland), N. Ahmad (India), M. Åmark (Sweden), J.T. Andrews (USA), J.W. Attig (USA), P.J. Barnett (Canada), B. Bergstrøm (Norway), G.S. Boulton (UK), P. Calkin (USA), J.J. Clague (Canada), L. Clayton (USA), R. Connell (UK), E. Derbyshire (UK), M. Deynoux (France), L.A. Dredge (Canada), E. Drozdowski (Poland), J. Ehlers (GFR), J.A. Elson (Canada), K. Erikkson (Sweden), E.B. Evenson (USA), V. Evzerov (USSR), M.M. Fenton (Canada), B. Frenzel (GFR), R.J. Fulton (Canada), G. Gillberg (Sweden), R.P. Goldthwait (USA), F. Grube (GFR), S. Haldorsen (Norway), M.J. Hambrey (UK), A. Hansel (USA), G.M. Haselton (USA), H. Heuberger (Austria), S.R. Hicock (Canada), A. Hillefors (Sweden), D. van Husen (Austria), W.H. Johnson (USA), P.F. Karrow (Canada), L.K. Kauranne (Finland), T.J. Kemmis (USA), R.W. Klassen (Canada), J. Krüger (Denmark), R. Kujansuu (Finland), E. Lagerlund (Sweden), P. La Salle (Canada), Y.A. Lavrushin (USSR), J. Lundqvist (Sweden), I. Marcussen (Denmark), C.L. Matsch (USA), R.W. May (Canada), A.M. McCabe (Ireland), I.C. McKellar (New Zealand), J.J.M. van der Meer (Netherlands), J. Menzies (Canada), K.D. Meyer (GFR), D.M. Mickelson (USA), H.H. Mills (USA), E.H. Muller (USA), W. Niewiarowski (Poland), S. Occhietti (Canada), A. Olszewski (Poland), G. Prichonnet (Canada), J. Rabassa (Argentina), M. Rappol (Netherlands), A. Raukas (USSR), B. Ringberg (Sweden), S. Rubulis (Argentina), E.V. Rukhina (USSR), H. Ruszczyńska-Szenajch (Poland), M. Ružicka (Czechoslovakia), M. Saarnisto (Finland), R. Souchez (Belgium), Ch. Schlüchter (Switzerland), G. Seret (Belgium), D. Serrat (Spain), J. Shaw (Canada), W.W. Shilts (Canada), S. Sjøring (Denmark), J.L. Sollid (Norway), W. Stankowski (Poland), H.J. Stephan (GFR), R. Sutinen (Finland), V.D. Tarnogradskiy (USSR), D.G. Vanderveer (Canada), K. Virkkala (Finland), T.O. Vorren (Norway), W.P. Warren (Ireland).

I am grateful to the Natural Sciences and Engineering Research Council of Canada and the International Union for Quaternary Research (INQUA) for supporting the preparation of this paper, to Mark McCrae and Hilkka Saastamoinen for drafting the diagrams and Stephan R. Hicock and Matti Saarnisto for helpful comments on the initial version of the paper.

REFERENCES

Åmark, M. 1985. Glacial tectonics and deposition of stratified drift during formation of tills beneath an active glacier – example from Skåne, southern Sweden. Boreas 15: 155-171.

Ashley, G.M., J. Shaw and N.D. Smith (eds.) 1985. Glacial sedimentary environments. SEPM short course No. 16. Society of Economic Paleontologists and Mineralogists, Tulsa OK.

Barnett, P.J. 1987. Quaternary stratigraphy and sedimentology, north-central shore Lake Erie, Ontario, Canada. Ph.D. thesis, University of Waterloo, Ontario, Canada.

Barnett, P.J. and R.I. Kelly 1987. XIIth INQUA Congress field excursion A-11. Quaternary history of southern Ontario. National Research Council of Canada.

Böhm, A. von. 1901. Geschichte der Moränenkunde. Abhandlungen der K.K. Geographischen Gesellschaft in Wien, III-4. Wien: R. Lechner.

Boulton, G.S. 1976. A genetic classification of tills and criteria for distinguishing tills of different origin. In W. Stankowski (ed.), Till, its genesis and diagenesis. Uniwersytet im. Adama Mickiewicza w Poznaniu, Seria Geografia 12: 65-80.

Boulton, G.S. 1980. Classification of till. Quaternary Newsletter 31: 1-12.

Boulton, G.S. 1987. A theory of drumlin formation by subglacial sediment deformation. In J. Menzies and J. Rose (eds.), Drumlin Symposium: 25-80. Rotterdam: Balkema.

Boulton, G.S. and M. Deynoux 1981. Sedimentation in glacial environments and the identification of tills and tillites in ancient sedimentary sequences. Precambrian Research 15: 397-420.

Boulton, G.S. and R.C.A. Hindemarsh 1987. Sediment deformation beneath glaciers: rheology and geological consequences. Journal of Geophysical Research 92: 9059-9082.

Boulton, G.S. and A.S. Jones 1979. Stability of temperate ice sheets resting on beds of deformable sediment. Journal of Glaciology 24: 29-43.

Brodzikowski, K. and A.J. van Loon 1987. A system classification of glacial and periglacial environments, facies and deposits. Earth-Science Reviews 24: 297-381.

Braun, A.F., R. German and M. Mader 1976. Der Beitrag der Sedimentanalyse zur Quartärstratigraphie. Bezirksstelle der Naturschafts Landschaftspflege Tübingen, Mitteilung 4. Tübingen.

Croot, D.G. 1987. Glacio-tectonic structures: a mesoscale model of thin-skinned thrust sheets? Journal of Structural Geology 9: 797-808.

Dreimanis, A. 1976. Tills, their origin and properties. In R.F. Legget (ed.), Glacial till. The Royal Society of Canada Special Publication 12: 11-49.

Dreimanis, A. 1979. The problems of waterlain tills. In Ch. Schlüchter (ed.), Moraines and varves: 167-177. Rotterdam: Balkema.

Dreimanis, A. 1982. Work group (1) – Genetic classification of tills and criteria for their differentiation: Progress reports on activities 1977-1982, and definitions of glaciogenic terms. In Ch. Schlüchter (ed.), INQUA Commission on genesis and lithology of Quaternary deposits, Report of activities 1977-1982: 12-31. Zürich: ETH.

Dreimanis, A. 1983. Penecontemporaneous partial disaggregation and/or resedimentation during the formation and deposition of subglacial till. Acta Geologica Hispanica 18: 153-160.

Dreimanis, A. 1987. Genetic complexity of a subaquatic till tongue at Port Talbot, Ontario, Canada. In R. Kujansuu & M. Saarnisto (eds.), INQUA Till Symposium, Finland 1985. Geological Survey of Finland, Special Paper 3: 23-38.

Dreimanis, A. 1989. Tills, their genetic terminology and classification. In R.P. Goldthwait and C.L. Matsch (eds.), Genetic classification of glacigenic deposits: 15-81. Rotterdam. Balkema.

Dreimanis, A., J.P. Hamilton and P.E. Kelly 1987. Complex subglacial sedimentation of Catfish Creek till at Bradtville, Ontario, Canada. In J.J.M. van der Meer (ed.), Tills and glaciotectonics: 78-87. Rotterdam: Balkema.

Dreimanis, A. and J. Lundqvist 1984. What should be called till? In L.K. Königsson (ed.), Ten years of Nordic till research. Striae 20: 5-10.

Dreimanis, A. and Ch. Schlüchter 1985. Field criteria for the recognition of till or tillite. Paleogeography, Paleoclimatology, Paleoecology 51: 7-14.

Drewry, D. 1986. Glacial geologic processes. London: Edward Arnold.

Elson, J.A. 1961. The geology of tills. In. E. Penner and J. Butler (eds.), Proceed. 14th Canad. Soil Mechanics Confer., National Research Council of Canada, Associate Committee on Soil and Snow Mechanics. Technical Memorandum 69: 5-36.

Elson, J.A. 1989. Comment on glacitectonite, deformation till, and comminution till. In R.P. Goldthwait and C.L. Matsch (eds.), Genetic classifications of glacigenic deposits: 82-85. Rotterdam: Balkema.

Evenson, E.B. and J.M. Clinch 1987. Debris transport mechanisms at active Alpine glacier margins: Alaskan case studies. In R. Kujansuu and M. Saarnisto (eds.), INQUA Till Symposium Finland 1985. Geological Survey of Finland Special Paper 3: 111-136.

Eyles, N. 1979. Facies of supraglacial sedimentation on Icelandic and Alpine temperate glaciers. Canadian Journal of Earth Sciences 16: 1341-1361.

Eyles, N., J.A. Sladen and S. Gilroy 1982. A depositonal model for stratigraphic complexes and facies superimposition in lodgement tills. Boreas 11: 317-333.

Geddes, R.S. 1982. The Vixen Lake indicator train, Northern Saskatchewan. In P.H. Davenport (ed.), Prospecting in areas of glaciated terrain – 1982. Canadian Institute of Mining and Metallurgy, Geology Division: 264-283.

Gibbard, P. 1980. The origin of stratified Catfish Creek Till by basal melting. Boreas 9: 71-85.

Gillberg, G. 1977. Redeposition: a process of till formation. Geologiska Föreningens i Stockholm Förhandlingar 99: 246-253.

Gravenor, C.P., V. von Brunn and A. Dreimanis 1984. Nature and classification of waterlain glaciogenic sediments, exemplified by Pleistocene, Late Paleozoic and Late Precambrian deposits. Earth Science Reviews 20: 105-166.

Grube, F. and Th. Vollmer 1985. Der geologische Bau pleistozäner Inlandgletschersediments Norddeutschlands. Bulletin of Geological Society of Denmark 34: 13-25.

Hansel, A.K., W.H. Johnson and B.J. Socha 1987. Sedimentological characteristics and genesis of basal tills at Wedron, Illinois. In R. Kujansuu and M. Saarnisto (eds.), INQUA Till Symposium, Finland 1985. Geological Survey of Finland Special Paper 3: 11-21.

Hughes, T.J. 1981. Numerical reconstruction of paleo-ice sheets. In G.H. Denton and T.J. Hughes The last great ice sheets: 221-261. New York: John Wiley & Sons.

Husen, D. van 1981. Die Ostalpen in den Eiszeiten. Wien: Geologische Bundesanstaltung.

Lawson, D.E. 1979. Sedimentological analysis of the western terminus region of the Matanuska Glacier, Alaska. Cold Region Research and Engineering Laboratory Report 79-9.

Lawson, D.E. 1989. Glacigenic resedimentation: classification concepts and application to mass-movement processes and deposits. In R.P. Goldthwait and C.L. Matsch (eds.), Genetic classification of glacigenic deposits: 144-166. Rotterdam: Balkema.

Meer, J.J.M. van der 1982. The Fribourg area, Switzerland. A study of Quaternary geology and soil development. Publicaties van het Fysisch-Geografisch en Bodemkundig Laboratorium van de Universiteit van Amsterdam 32.

Mills, H.H. 1977. Textural characteristics of drift from some representative Cordilleran glaciers. Geological Society of America Bulletin 88: 1135-1143.

Muller, E. 1983. Till genesis and the glacier sole. In E.B. Evenson, Ch. Schlüchter and J. Rabassa (eds.), Tills and related deposits: 19-22. Rotterdam: Balkema.

Parkin, G.W. and S.R. Hicock 1989. Sedimentology of a Pleistocene glacigenic diamicton sequence near Campbell River, Vancouver Island, British Columbia. In R.P. Goldthwait and C.L. Matsch (eds.), Genetic classification of glacigenic deposits: 93-112. Rotterdam: Balkema.

Pessl, F. and J.E. Frederick 1981. Sediment source for melt-water deposits. Annals of Glaciology 2: 92-96.

Rabassa, J., S. Rubulis and J. Suarez 1979. Rate of formation and sedimentology of (1976-1978) push-moraines; Frias Glacier, Mount Tronador (41°10'S, 71°53'W), Argentina. In Ch. Schlüchter (ed.), Moraines and varves: 65-79. Rotterdam: Balkema.

Rappol, M. 1983. Glacigenic properties of till. Studies in glacial sedimentology from the Allgäu Alps and The Netherlands. Publicaties van het Fysisch-Geografisch and Bodemkundig Laboratorium van de Universiteit van Amsterdam 34.

Ruszczyńska-Szenajch, H. 1983. Lodgement tills and syndepositional glacitectonic processes related to subglacial thermal and hydrologic conditions. In E.B. Evenson, Ch. Schlüchter and J. Rabassa (eds.), Tills and related deposits: 113-117. Rotterdam: Balkema.

Sharp, R.P. 1949. Studies of superglacial debris on valley glaciers. American Journal of Science 247: 289-315.

Shaw, J. 1977. Tills deposited in arid polar environments. Canadian Journal of Earth Sciences 14: 1239-1245.

Shaw, J. 1985. Subglacial and ice marginal environments. In G.M. Ashley, J. Shaw and H.D. Smith (eds.), Glacial Sedimentary Environments. SEPM Short Course 16. Tulsa: Society of Economic Paleontologists and Mineralogists: 7-84.

Shaw, J. 1987. Glacial sedimentary processes and environmental reconstruction based on lithofacies. Sedimentology 34: 105-116.

Shaw, J. 1989. Sublimation till. In R.P. Goldthwait and C.L. Matsch (eds.), Genetic classification of glacigenic deposits: 138-139. Rotterdam: Balkema.

Stephens, G.C., E.B. Evenson, R.B. Tripp and D. Detra 1983. Active alpine glaciers as a tool for bedrock mapping and mineral exploration: a case study from Trident Glacier, Alaska. In E.B. Evenson, Ch. Schlüchter and J. Rabassa (eds.), Tills and related deposits: 195-204. Rotterdam: Balkema.

Sugden, D.E. and B.S. John 1976. Glaciers and landscape, a geomorphologic approach. London: Edward Arnold.

Tilas, D. 1740. Tankar om malmletande i anledning af löse gråstenar. Kongl. Svenska Vetenskaps-Akademiens Handlingar. I. 1739-1740: 190-193.

Vorren, T.O. 1977. Grain-size distribution and grain-size parameters of different till types on Hardangervidda, south Norway. Boreas 6: 219-227.

Warren, W.P. 1989. Protalus till. In R.P. Goldthwait and C.L. Matsch (eds.), Genetic classification of glacigenic deposits: 145–146. Rotterdam: Balkema.

APPENDIX

A-1. *Introduction*

During the years 1978-1980 the Till Work Group of the INQUA Commission on Genesis and Lithology of Quaternary Deposits discussed the descriptive criteria of the two groups of tills: basal till and ablation till. The discussions began by correspondence, with replies to the Till Work Group circulars; they continued at two field conferences – in Switzerland and Norway.

The participants of the Finse meeting in Norway, August 1979, proposed to use the term 'supraglacial till' instead of 'ablation till', but the opinions about the useage of 'basal till' or 'subglacial till' were evenly split. The majority of written replies to a questionnaire sent out on 20 November 1979, preferred the terms 'ablation till' and 'basal till'. However, the participants of a regional meeting at Keele, U.K. in December 1979, proposed that the term 'ablation till' '...should not form part of a recommended classification' (Boulton 1980: 3). Because of these widespread differences in opinions on the dominantly positional terminology, and a prevalent opinion among many Till Work Group members, that a genetic classification based upon the process of deposition, such as lodgement, melt-out, etc., would be more desireable, further discussion of the positional varieties of till was shelved temporarily in 1980.

However, detailed investigations of the variety of facies in a single unit of till (see references in section: Simultaneous deposition at several levels, and also in Dreimanis 1989: Chapter 10) during the 80-ies demonstrated that several of the depositional genetic varieties commonly interdigitate or alternate both laterally and vertically on short distances and, in some till sections it is possible to identify the positional varieties only such as 'subglacial or basal till', and 'supraglacial or surface ablation till', as mappable units.

The discussions on descriptive properties on the above two positional groups of till were resumed by correspondence during the 1983-86 period, but their results including those of 1978-80, were only briefly summarized in Dreimanis (1989). Since the genetic classification of tills by their position of deposition is utilized particularly by prospectors, the present report is probably a suitable place to include the descriptive criteria of supraglacial and subglacial tills, by compiling those opinions of the 85 members of the INQUA Till Work Group (listed in: Acknowledgements), that have been expressed during the 1978-86 period and that were in agreement with the observations of the author and with data published in the papers referred to in this report. Before listing the descriptive properties, the derivation, transport and depositional genetic varieties for each of two groups will be summarized.

## A-2. *Subglacial or basal till*

### A-2.1. *Derivation and transport*
Basal till originates from subglacially derived debris. They are transported, prior to deposition, either (1) by subsole drag for a short distance, or (2) in the basal zone of ice, also for a short distance in warm-based glaciers with extending flow, or variable to long distance, depending upon other combinations of the flow- and temperature-regime at the base of glacier ice. If any resedimentation by gravity flow or squeeze flow occurs, its transport is for short distances only. Boulton (1987: 42) suggests that his massive fine-grained A-horizon produced by subglacial deformation and considered by him to be deformation till, might be moved considerable distance, depending upon the time of transport. However, these great distances have not been proven by any case studies.

### A-2.2. *Depositional varieties*
Deformation till and squeeze flowtill, lodgement till, basal melt-out till, and subglacial gravity flowtill, usually in several combinations as sub-units. Most of them are primary tills.

### A-2.3. *Descriptive properties*
*Position and basal contact.* Subglacial till is usually the lowermost part of a stratigraphic till unit, on erosional contact with substratum.

*Lateral extent and thickness.* Usually a laterally extensive mappable unit; thickness – variable, one to more than ten metres, but the lateral change of thickness is gradual.

*Surface expression, landforms.* Fluted, drumlinized or gently rolling in till plains or ground moraine landscapes. Also as low transverse ridges ('minor moraines') formed subglacially, and on the proximal sides of some end moraine ridges.

*Structure folding, jointing.* Usually described as massive and matrix-supported, but, on close examination, a variety of consistently oriented macro- and micro-structures are visible, such as: (a) fissility or foliation, particularly in the upper part of the till unit, (b) shear planes, subhorizontal or rising downglacier, most common in the lower part, (c) various glacitectonic deformations, with the crests of recumbent anticlines attenuated downglacier, also most common in the lower part, (d) vertical joint systems, bysected by the stress direction, and transverse joints steeply dipping downglacier. The orientation of all the above deformation structures is related to the stress applied by the moving glacier and, therefore, they are consistently oriented for some distance laterally. Where resedimentation has occured by squeeze flow or gravity flow in subglacial or englacial cavities, particularly on lee sides of bedrock protrusions, flow structures, unrelated to the glacial movement, have developped. Also, flowtills are interbedded with stratified sediments or contain lenses of stratified sediments.

*Clasts and their surface marks.* Subrounded to subangular shapes dominate, except for deformation till, if it is formed of local bedrock: then its fragments are angular. Some of the more distantly transported clasts are bullet shaped, some are facetted, also sheared. Pavements or concentrations of clasts ('boulder pavements') occur particularly in the lower part and at levels where deposition of basal till was temporarily interrupted by local erosion. The tops and often also the bottoms of clasts in intra-till pavements are consistently striated along the direction of the local glacial movement at the time of their formation.

*Orientation of clasts ('till fabric').* Usually well developed with parallel modes dominating; transverse modes are also present, particularly in deformed parts of till; fabric maxima are consistent within site and between sites, except for local deviations around large boulders or bedrock protrusions, or if the glacial movement had changed during the deposition of till and caused also some re-orientation of the previous fabric. If the glacial movement remains unchanged the fabric modes are in good agreement with the orientation of glacitectonic structures and the main fabric mode is parallel or transverse to the dominant sets of striae on clasts and on the underlying bedrock. Exception: local fabrics unrelated to the glacial movement have developed in the flowtills.

*Grain-size composition.* Usually a diamicton, containing clasts of various sizes. Grain-size composition depends greatly upon the lithology and grain-size composition of the substrata upglacier, and the distance and process of transport of them. Each basal till layer, except for those rich in megaclasts or glacitectonic deformations, has a relatively constant laterally traceable grain size composition; subhorizontal lenses of differing grain size composition are present. Sorting is usually poor or very poor, except for basal tills (deformation till) that consist mainly of reworked sorted material; skewness has a nearly symmetrical distribution.

*Lithologic composition.* Most rock fragments are of local derivation, particularly in the lower part of basal till; distantly transported clasts are more common or may even dominate in the upper part of basal till if it has been formed of englacially transported debris. Till matrix usually contains both local and distant minerals (mainly 'rock flour', a product of glacial abrasion and crushing); the basal part particularly deformation till, is enriched in local minerals. The lithologic composition of each genetic subunit is laterally consistent for some distance, if the substratum material does not change radically.

*Consolidation, permeability, density.* Commonly overconsolidated, except for deformation till and flowtills, if there was adequate subglacial drainage during deposition. Permeability is low in lodgement till, variable in other varieties of subglacial till. Bulk density is highest in lodgement till, lowest in flowtills and local tills.

*Diagnostic properties.* Basal till usually occurs in the lower part of a stratigraphic till unit, and is also its main component in those areas where streamlined landforms suggest deposition under actively moving glacier ice. The lower contact is usually erosional. The orientations of fabric, glacidynamic deformation structures, major fracture sets, fissility, parallel sets of striae on clasts, and glacial sole marks are all oriented consistently with the direction of stress applied by the actively moving glacier during the formation of basal till, or basal transport in the case of till deposited by basal melt-out. Local exceptions are subglacial and englacial flowtills, deposited on the lee side of bedrock protrusions, or in other cavities. Texturally the till is commonly a matrix-supported diamicton with a multimodal particle size distribution, except for lowermost deformation till subunit that texturally and lithologically resembles the deformed substratal material. Lithologic composition is usually laterally consistent for a considerable distance. The subglacial till's lower part of clasts are influenced more by the subsole material (both bedrock and soft sediments) than its main body and its matrix. Most clasts are subangular to subrounded, bearing glacial abrasion marks, except for any local till, commonly called local moraine, a variety of deformation till, at its base, where angular clasts dominate.

### A-3. *Supraglacial or surface ablation till*

Since surface ablation is the main process of the formation and deposition of this variety of till, its most popural name is 'ablation till'. If this term is used, it would be adviseable to call it 'surface ablation till'. As already discussed in section: Genetic classification of till..., two derivation- and transport-related varieties of supraglacial till are distinguished: (a) exogenous till formed of extraglacial debris supplied by mass movements to the glacier surface, and (b) basally derived supraglacial till. In the following discussion of supraglacial tills an asterix (*) will be placed along the left side of the column, where any specific characteristics of exogenous till will be given.

#### A-3.1. *Derivation and transport*
\* On mountain glaciers and along nunataks protruding through ice sheets, supraglacial debris derive extraglacially by falling and/or sliding upon the glacier surface from valley sides or nunataks. Wind also contributes some fine-grained debris. All supraglacial debris are transported passively on glacier ice, in snow and firn, or in the upper part of glacier ice, without being affected by glacial abrasion or comminution.

In all glaciers supraglacial debris derive also subglacially, reaching the glacier surface via englacial transport, by upward movement of debris layers from the base of the glacier, mainly under compressive flow regime. They undergo glacial abrasion and comminution.

Prospectors would be particularly interested whether there is any difference in the length of glacial transport of debris among the two varieties of supraglacial till. The supraglacial transport of exogenous debris may be of any length (Stephens et al. 1983). As for the basally derived debris, usually two extremes have been encountered, and occasionally even in a single section: (a) a very short transport by compressive flow, and (b) any distance of transport, mainly englacially, by extending flow (Dreimanis 1976: 27; Geddes 1982).

### A-3.2. *Deposition*

The formation of supraglacial till may begin already on glacier ice, by concentration of supraglacial debris due to melting of snow, firn and ice (supraglacial melt-out till) or sublimation in very cold and dry climate (sublimation till), also by mass movements of supraglacial debris downslope to a lower ice surface (supraglacial flowtill), with minor participation of meltwater. The deposition is completed by melting of the underlying ice or by mass movements of supraglacial debris or supraglacial melt-out till towards the land surface or into water adjacent to glacier ice (ice-marginal flowtill, glacioterrestrial, or glacioaquatic).

Deposition typically occurs on and against stagnant ice but it takes place also on and along active glaciers. The till deposits of retreating glaciers and those deposited during the maximum extent of glacial termini are more likely to be preserved than those of advancing glaciers.

### A-3.3. *Descriptive properties*

*Position and basal contact.* Supraglacial till is the uppermost glacial sediment in a non-aquatic glacial facies association. Flowtill is commonly interbedded or interdigitated with glacioaquatic sediments. Basal contact of the supraglacial melt-out till is usually planar and concordant with the top of basal melt-out till. Flowtills commonly fill depressions or shallow channels; their basal contacts vary from concordant to erosional.

Note: washing and frost action may cause the surface of coarse textured subglacial till to resemble supraglacial, particularly exogenous till.

*Lateral extent and thickness.* Extremely variable, even in the same unit. It may be from a few centimetres or a boulder lag to several tens of metres thick. Thickest supraglacial tills are in areas of mountain glaciation, particularly in lateral moraines and in the areas of glacial stagnation. Thin but laterally extensive ablation blanket covers glacioterrestrial basal till landforms deposited by active, but retreating glacier.

*Surface expression, landforms.* The surface expression varies, depending upon the dominant type of glacial flow, the climatic conditions which influenced the ablation, the amount of supraglacially and englacially transported material and the morphology of the ground surface upon which the till was deposited. Supraglacial till is mainly found in hummocky disintegration moraines, in lateral and medial moraines and in transverse ice-marginal ridges (dump moraines).

*Structure, folding, faulting.* Variable structure – either massive, or interbedded diamictons and sorted sediments, or diamictons containing lenses of sorted sediments. Flowtills may display a variety of flow structures and soft sediment deformation structures; their orientation is unrelated to glacial movement, but related to local non-glacial stresses, and it varies from place to place. Melt-out till may have preserved glaciodynamic structures inherited from glacier ice. Melting of underlying ice produces local sagging, gravity faulting or thrusting. Exogenous till is often clast supported.

*Clast shapes and their surface marks.* Angular clasts are more abundant in surface ablation tills than in other tills, due to (a) presence of extraglacially derived and supraglacially transported talus material in exogenous till of mountain glaciers, or (b) frost shattering on the glacier surface. Where clasts have been basally derived and transported first, subangular and subrounded striated stones are also present; in areas where glaciers had incorporated proglacial outwash or inwash gravels, rounded pebbles are abundant.

*Orientation of clasts ('till fabric').* Very variable: random to well defined. Three types of clast-orientation fabrics occur: (a) fabrics related to englacial movement in melt-out or sublimation tills; (b)

\* fabrics related to mass movement in flow tills; (c) fabrics related to mass movements, especially by landslides, on the surface of ice; also combinations of any of them. Fabrics (b) and (c) are unrelated to glacial movement and their modes vary from place to place.

*Grain size composition.* More variable than in related basal tills, usually with a higher proportion of coarser grained material: more sand and clasts than in the coarse textured related basal till, more silt
\* than in the related clayey-silty subglacial till. Particularly coarse are exogenous tills, some of them having modes in the granule to cobble sizes. Sorting commonly is better than in basal tills due to surface winnowing of fines and to lesser production of fines by clast-to-clast contact or due to admixture of sorted material.

*Lithologic composition.* More variable than in related basal till. Composition is likely to differ from underlying basal till depending on bedrock topography, and the thermal and dynamic regimes of the glacier. In ice marginal areas with strong compressive flow supraglacial till may be enriched in 'local' clasts. In areas of extending flow supraglacial till is richer in far-travelled debris. In some cases the lithologic differences between the two facies are non-existent, slight, or difficult to define.
\* Exogenous supraglacial till may be lithologically different from basally derived supraglacial till, if their sources were different.

*Geotechnical properties.* Permeability is commonly high in surface ablation tills, except in clay rich
\* varieties where permeability could be primarily controlled by desiccation cracks. Exogenous tills are usually quite porous and normally consolidated. Their coarse texture gives them a low bulk density and, consequently, a loose appearance.

*Diagnostic criteria.* Supraglacial tills are in the uppermost part of a stratigraphic till unit. They occur in variable thickness mainly in hummocky disintegration moraines, lateral, medial and dump end moraines. They form a thin blanket over glacioterrestrial basal till that is deposited by an active but retreating glacier.
\* In areas of mountain glaciation angular extraglacial debris dominate in supraglacial till; it is coarse textured and loose. The exogenous till of medial moraines is lithologically closely related to their extraglacial derivation places.

Basally derived supraglacial till is texturally and lithologically more variable than the related underlying basal till. Texturally it is coarser, sandier and richer in clasts than the related sandy-silty basal till, and more silty than the related silty-clayey basal till. Lithologically it either reflects distant or local bedrock, depending upon the predepositional transport: by extending flow in the first case, and by compressive in the second one.

# CHAPTER 4

# Glacial morphology as an indicator of the direction of glacial transport

JAN LUNDQVIST

*Department of Quaternary Research, University of Stockholm, Sweden*

## INTRODUCTION

The locating of ore bodies by means of finds of glacially transported boulders containing mineralized rocks has been of great importance since older times (Tilas 1740), but tracing of the sources of such boulders sometimes poses considerable problems. If the boulders are frequent and form a boulder train, a reasonably exact picture of the distribution is obtained, pointing to the source, the ore body. Very often, however, only scattered boulders are found. An instructive example has been described by G. Lundqvist (1947). In such a case the direction and distance of transport must be estimated with sufficient accuracy to enable a search for further boulders to be continued. The following pages deal with the reconstruction of the direction of glacial transport by means of morphological elements.

The elements concerned vary greatly in magnitude. The major landforms constitute one extremity, while the opposite extremity is represented by glacial striae, down to the tiniest scratches. In between we find landforms representing glacial deposits, particularly moraine forms.

## GLACIAL STRIAE AND ROCK SCULPTURE

The most exact, and also most widely used, method for determining the direction of glacial transport is to study glacial striae and other small-scale structures on outcrops of the bedrock. The outcrops in a formerly glaciated terrain are mostly rounded and shaped by the glacial erosion. Earlier there was a widespread opinion that these forms were entirely created by the glacier from a flat or water-eroded land surface. Recent investigations (see the Symposium volume Fennia 163: 2, 1985, and Lidmar-Bergström 1982) indicate that these forms may be controlled to a great extent by differential deep weathering of the bedrock and just modified by glacial activity. The result is often more or less well developed stoss and lee sides on the outcrop – roches moutonnées. If the ice movement happens to have been consistent throughout the glaciation, or during several glaciations, these forms are most pronounced and give a fairly good idea of the direction of this movement. Where ice movements have shifted, and especially in areas close to an ice divide, the picture is less clear.

A more precise picture is given by the glacial striae, which enable the direction to be determined with great accuracy. Very often more than one direction can be traced,

Figure 1. Glacial striae indicating at least five stages of ice movement. The coarse set from 80° (from upper right to lower left) is the oldest, and is crossed by intermediate sets from 195°, 140° and 100°. A finer set from 330° touching only the top parts of the outcrop is youngest. East of Sädvaluspa, Swedish Lapland. Photo J. Lundqvist 1971.

represented by different sets of striae, in which case determination of the age relations between these sets, that is, the pattern of shifting ice movements, is essential. The most important and unequivocal method of doing this is by means of the positions of the striae in relation to each other upon the outcrop (Fig. 1). A set located in a lee-side position relative to another, is usually older, a set just touching the top parts of the outcrop will have been formed by the youngest movement, and a set preserved only in depressions and other low positions may be interpreted as being older.

This method of analyzing the interrelations between striae is more accurate and easier to apply than the direct study of cross-striae, where very close examination by means of magnifying lenses or the like is necessary to reveal whether one set of striae is formed in the bottom of the other striae. Otherwise a set of finer striae will usually be interpreted as younger than a coarser set.

Other features of glacial erosion or pressure, such as parabolic shattermarks, crescentic fractures etc. often occur together with striae. These give a general idea of the direction of the pressure but not as exact as the striae. If the stoss and lee sides are not clearly developed they may be useful for determining these parameters.

The interpretation of these small forms presents many problems. For instance, they indicate the movement of the glacier, but give little information on the duration of this movement or its absolute age. An older set of striae may represent a phase just slightly older than the deglaciation, or an older glaciation.

Another complication is the influence of the local topography, and even the forms of the outcrop itself. Striae may turn around a steep outcrop or be deviated by furrows and crevasses. If the surrounding higher parts of the outcrop are eroded down, such local deviations will scarcely be identified as such and may be misinterpreted as being of regional significance.

For these reasons a very careful study of the features is necessary for the interpretation of directions of ice movement. Ljungner (1930) developed methods for analysing glacial rock sculpture which even allow the glaciological conditions and the duration of different phases of movement to be estimated. Without such careful analysis the ice movements identified tell us nothing about the key problem in this context, the glacial transport of significant boulders or finer debris.

## LARGE-SCALE LANDFORMS

Roches moutonnées with their striae represent small-scale erosion and scouring of the land surface, but corresponding features are also found on a very large scale. Larger hills and even mountains may be sculptured like roches moutonnées, thus indicating the direction of ice flow. These are mostly combined with accumulations which mask their lee side. Thus the lee side is often apparently as smoothly shaped as the upstream side. Because the stoss side is usually formed of bedrock and the lee side of till or other glacial sediments, there is nevertheless no problem in identifying the direction of flow.

A false roches moutonnées sculpture may sometimes complicate the picture locally. Where a rounded mountain or hill is affected by a deeper incised valley, there may be a false, steep lee side facing the valley. This gives a misleading impression of the ice flow, but a consideration of the regional picture of landforms will immediately reveal the true nature of these features.

Large landforms cannot have been shaped during one single glaciation, but require repeated glaciations with a consistent flow pattern. It may be inferred that the earliest glaciations affecting a deeply weathered terrain are the most important for this kind of sculpture. For this reason large landforms are less informative for identifying the glacial transport of specific objects, merely giving an indication of a dominant flow direction. In Scandinavia we find these forms indicating a flow from the mountain range towards the lowland, a pattern which was most probably repeated at some time during each glaciation.

## MORAINE FORMS

More informative are the smaller landforms, shaped by the accumulation of glacial debris in the form of till. It is true that these may partly be erosional, which causes complications, as discussed in the following, but in general they are important as indicators of glacial transport. We may distinquish a few groups of forms, essentially characterized as radial, transverse, or zones with a variety of forms.

Figure 2. Drumlins are protracted in the direction of ice flow and serve as a good indicator of this direction. The interior, however, may have been transported in a different direction. Adelaide Peninsula, NW Territories, Canada. Photo J. Lundqvist 1984.

### Radial moraine forms

The radial forms are elongated in the direction of flow, that is, radially in relation to the ice sheet. Essentially they are accumulations, but erosion and scouring of older surfaces of till or other unlithified sediments will give similar forms, as discussed below. A continuous series of landforms may be identified, starting with roches moutonnées and ending with purely glacial deposits.

The simplest form in this series is an accumulation on the lee side of an obstacle. These vary in magnitude within wide limits. A large landform such as a mountain with some accumulation on its distal end is an extreme variety. Most common, and most important from the present point of view, is the crag-and-tail form, i.e. a rock with a tail of glacial sediments pointing downstream. Similar forms commonly develop on the lee side of individual boulders, although they are easily destroyed by solifluction and other ground processes and are therefore seen mainly close to recent glaciers.

The opposite type of landform is represented by precrags, which are formed in favourable positions by the packing of basal till against obstacles of varying size (cf. Gillberg 1976).

The combination of a precrag and a tail gives a landform consisting of a core of bedrock smoothened by the deposition of glacial sediments and elongated in the direction of flow. All transitions exist, from bedrock hills – and roches moutonnées – through smoothened hills in which the bedrock is clearly visible to elongated hills with glacial deposits completely covering the core.

Figure 3. Flutings on till surfaces are a very good indicator of the last main ice flow over the area. The margin of Langjökull, Iceland. Photo J. Lundqvist 1962.

The latter landforms approach drumlins, which are smoothly shaped, streamlined or whaleback forms – accumulations of glacial deposits (Fig. 2). Even a true drumlin was probably initiated by some obstacle, and may contain a core of bedrock or older deposits, even though this is not observed. For practical purposes it is best to consider all such whaleback forms drumlins, while those consisting of bedrock to a considerable and visible extent may be called drumlinoids.

A perfect drumlin gives direct information on the direction of ice flow, but closer study will often reveal complicating irregularities. The direction of flow may have shifted, giving the drumlin a curved shape. A drumlin may also be affected and modified by different flows, sometimes at considerable angles one to the other. This is important from the point of view of glacial transport, for it is necessary to determine to which flow stage the glacial sediment corresponds.

Fluted moraines are minor but important landforms related to drumlins (Fig. 3). They

Figure 4. The ridges of Rogen moraine give an approximate idea of the direction of ice flow. The flow was at right angles to the ridges, with the convex sides of their individual, shorter constituents pointing upstream. In this case, near Lake Rogen, Härjedalen, Sweden, the flow was towards the viewer, from the upper left. Photo J. Lundqvist 1963.

may look like striated till surfaces – and in some instances probably are – but they are actually depositional forms. The relief does not exceed a few centimetres or decimetres, but the length of an individual flute may amount to hundreds of metres or even several kilometres.

Because of their low relief, flutings are rarely identified on the ground, except on the barren ground in front of recent glaciers. They are often to be seen in air photographs, however. In this case there may be some risk of misinterpretation caused by the shadows of trees, underlying bedrock structures, and even by effects resulting from the photographic processing.

Allowing for such complications, flutings are of the greatest importance for the determination of flow direction. Such a fragile structure seldom survives an ice flow in a different direction, and consequently it gives a good picture of the last flow. This is especially important because flutings are most common on wide till plains with no outcrops to carry striae.

Opinions differ concerning the process of formation of both drumlins (Glückert 1973, Gillberg 1976) and flutes (cf. Boulton 1976, Åmark 1980), although this is less relevant for the practical purpose of identifying directions of transport. Other complications exist, as discussed below.

The radial forms useful in this context are all subglacially formed. Their morphology is smooth and regular and the entire terrain is clearly shaped by moving ice. These forms must be distinguished from those which represent fillings in open or closed crevasses in stagnant ice, which are much more irregular in all respects. They can incidentally be extended in the direction of flow, but can just as well be elongated in any other direction. They are virtually useless for the determination of ice-flow directions.

*Transverse forms*

There are several moraine forms which run transverse to the direction of ice movement. Some of them are crevasse fillings, like the ones mentioned above, and are similarly useless for the determination of the direction of flow. They may give a rough idea of the direction, but this is too inaccurate for practical purposes. This should especially be noted in areas with De Geer moraines (Hoppe 1959), for if these are interpreted as end moraines, according to De Geer's (1940) original opinion, they can easily give a misleading picture. More probably these moraines should be considered mainly to be crevasse fillings (Hoppe 1957, Elson 1957, Strömberg 1965) like the ones mentioned above.

The features which can be used are the ones which form either along the ice margin or subglacially. The ice-margin deposits are moraines formed at the front of a glacier (end moraines, terminal moraines) or laterally along a glacier tongue. In the first case the moraines give a direct indication of the movement – at right angles to the front. The lateral moraines do not give a good picture of the direction but their pattern, especially together with frontal moraines, indicates the form of the ice and thereby the general movement. Mostly these forms have developed at local glaciers or glacier tongues extending from ice sheets, and therefore they are all of rather limited use. The flow pattern may just as easily be deduced from a general knowledge of the topography and extent of former ice sheets.

Rogen moraines and related forms are more useful in this respect. Rogen moraines form a landscape characterized by large ridges approximately at right angles to the flow (Fig. 4). The individual ridges are to a great extent composed of crescentic parts, each with its convex side turned upstream and horns pointing downstream (J. Lundqvist 1969, 1989). This does not apply to every ridge, but should be perceivable in all true Rogen moraine terrains. The crescent-shaped ridges can be considered incomplete drumlins in which the distal end is missing.

There is a gradual transition from drumlins to Rogen moraines, the latter being mainly located in basins in the terrain and the drumlins in convex positions. Transitions from drumlins to Rogen moraines can take place sideways as well as in the direction of flow, and upstream as well as downstream. This terrain-controlled distribution is clearly visible in Scandinavia (J. Lundqvist 1969).

The Rogen ridges also show other transitions to drumlinized moraines. Sometimes they are composed of small, complete drumlins aligned side by side, while in some areas the crests of the Rogen moraine ridges are fluted at right angles to the main ridge direction

(Wastenson 1969). These features, as well as a thorough morphological analysis of the ridges, can give a fairly accurate picture of the direction of ice flow. This is important, because Rogen moraines occur in positions where outcrops with striae are often rare.

### Zones of irregular moraine forms

End moraines were discussed above as giving an approximate idea of the general flow direction. In that context reference was made to ridges extending along, or parallel to, a former ice margin. The term end moraine, or simply moraine, often refers to a different type of formation. The moraine then consists of a broad zone of hummocky moraine often of very irregular form (cf. Alden 1918, Thwaites 1943, Clayton and Moran 1974, Nelson and Mickelson 1977). This represents the fringe zone of a glacier, where compressive flow caused up-transport of debris to the glacier surface (Boulton 1972, Eyles 1983). The debris was then deposited during the downwasting of the underlying ice to form a hummocky landscape of supraglacial till ('dead-ice moraine'). From the chronological point of view such a zone of hummocky moraine corresponds to a certain stage in the glaciation and is consequently comparable to an end moraine.

An end moraine of this type may be used as an approximate indication of the direction of glacial flow, like a ridge-shaped moraine, although individual hummocks and ridges within it give no directional information. For this purpose the regional context must be studied, e.g. by thorough mapping, or by interpretation from air photographs, and the eventual determination is just as approximate as that obtained from the study of single end moraine ridges, or even more so.

### GLACIOFLUVIAL FORMS

Glaciofluvial deposits are formed by meltwater from the ice either at and in front of the glacier margin or behind it, in crevasses or tunnels in the ice. They may give some general information about the direction of glacial flow, but this is very general and is mostly of no significance compared with moraine forms and striae. Proglacial (extramarginal) deposits are useless in this respect, but marginal deposits are compatible with end moraines, for end moraines are in fact very often composed of glaciofluvial sediments.

The deposits created behind the ice margin give a general idea of the flow pattern, but this is highly inaccurate, on a par with the information given by other crevasse fillings. Where eskers present an orientation which is not in agreement with the topography these may give some additional information about the dynamics of the glacier, but from the point of view of glacial transport this is of minor importance.

### COMPLICATIONS AND CONCLUSIONS

A number of features indicating the direction of ice flow have been discussed in the foregoing. A knowledge of this direction is of course of fundamental importance for the reconstruction of glacial transport, but we must not consider the two parameters to be identical. There are some complications.

The direction of flow, indicated in the ways discussed above, mostly represents only the last flow stage, especially in the case of finer, most exact features such as flutings, drumlins and Rogen moraines, and above all marginal formations. On the other hand, the very large forms often represent old flow stages, mostly several stages. Only the striae, where more than one set occurs, show the progression of different flow patterns. The striae themselves cannot be dated, however, and consequently we do not know whether they show the whole development or just changes in glacial movement near the ice margin.

The transport of glacial debris may have been influenced by several ice movements, and even if the debris has disintegrated during transport, individual boulders may survive several flow stages and consequently be transported along complicated trajectories. The evidence from landforms should therefore be filled out with other investigations. Studies of the till and its genesis may be very informative in indicating whether it is of local or distant origin. Its lithology gives general information, and closer investigation with attention paid to the glaciodynamics of the formation may be of great value (see Minell 1978).

Of major importance is the correlation between landforms and stratigraphy. Since some landforms may be at least partly the result of erosion, their internal composition could theoretically lack all correlation with the morphology. Drumlins are a good case in point. It is well known that drumlins may consist mainly of glaciofluvial and other water-laid sediments (Whittecar and Mickelson 1979), and even organic beds (Hillefors 1969), which are obviously older than the glacial stage responsible for the landform. Thus we may assume that the same applies to drumlins composed of till, although this is more difficult to prove.

As a consequence, landforms, although of great importance, should not be used alone to identify the direction of transport, but should always be combined with investigations of stratigraphy, lithology and till fabric, for instance. The relation between a stratigraphy unit containing boulders of interest from the point of view of prospecting and the local landforms must be established. This can often be done accurately enough when the boulders occur in till, and complementary studies of till genesis and glaciodynamics can sometimes give a reasonably good idea of the transport even of an isolated boulder, sufficient at least to promote the location of more boulders. Eventually a boulder train may be constructed in this way.

If a boulder train with a sufficiently dense population has been obtained, the landforms may be extremely useful for the exact location of the source, the ore body. At this stage a thorough analysis of the striae and landforms present is necessary. An excessively rough idea of the transport pattern, obtained from the boulder train alone, may lead to a waste of money through wrongly placed drilling sites. Final adjustment of the location according to analyses of the types discussed here may be important in this respect. A good case in point is described by Nilsson (1973).

REFERENCES

Alden, W.C., 1918. The Quaternary geology of southeastern Wisconsin, with a chapter on the older rock formations. U.S. Geological Survey Professional Paper 106. 356 pp.
Åmark, M., 1980. Glacial flutes at Isfallsglaciären, Tarfala, Swedish Lapland. Geologiska Föreningens i Stockholm Förhandlingar 102: 251-259.

Boulton, G.S., 1972. Modern Arctic glaciers as depositional models for former ice sheets. Journal of the Geological Society of London 128: 361-393.

Boulton, G.S., 1976. The origin of glacially fluted surfaces – observations and theory. Journal of Glaciology 17: 287-309.

Clayton, L. and S.R. Moran, 1974. A glacial process-form model. In: D.R. Coates (ed.), Glacial geomorphology. Publications in Geomorphology, State University, New York: 89-119.

De Geer, G., 1940. Geochronologia Suecica Principles. Kungliga Svenska Vetenskaps-Akademiens Handlingar 3: 18:6. 367 pp.

Elson, J.A., 1957. Origin of washboard moraines. Geological Society of America, Bulletin 68: 1721.

Eyles, N., 1983. Modern Icelandic glaciers as depositional models for 'hummocky moraine' in the Scottish Highlands. In: E.B. Evenson, Ch. Schlüchter and J. Rabassa (eds.) Tills and Related Deposits. A.A. Balkema, Rotterdam: 47-59.

Gillberg, G., 1976. Drumlins in southern Sweden. Bulletin of the Geological Institute, University of Uppsala, N.S. 6: 125-189.

Glückert, G., 1973. Two large drumlin fields in Central Finland. Fennia 120. 37 pp.

Hillefors, Å., 1969. Västsveriges glaciala historia och morfologi. Naturgeografiska studier. Meddelanden, Lunds universitets geografiska institution, Avhandling 60. 319 pp.

Hoppe, G., 1957. Problems of glacial morphology and the Ice Age. Geografiska Annaler A 39: 1-18.

Hoppe, G., 1959. Glacial morphology and inland ice recession in northern Sweden. Geografiska Annaler A 41: 193-212.

Lidmar-Bergström, K., 1982. Pre-Quaternary geomorphological evolution in southern Fennoscandia. Meddelanden, Lunds universitets geografiska institution, Avhandling 91. 202 pp.

Ljungner, E., 1930. Spaltentektonik und Morphologie der schwedischen Skagerrak-Küste. Bulletin of the Geological Institutes, University of Uppsala 21: 255-478.

Lundqvist, G., 1947. Ice-Movements and Boulder Trains in the Murjek-Ultevis Districts. Sveriges Geologiska Undersökning C 487: 80-91.

Lundqvist, J., 1969. Problems of the so-called Rogen moraine. Sveriges Geologiska Undersökning C 468. 32 pp.

Lundqvist, J., 1989. Rogen (ribbed) moraine – identification and possible origin. Sedimentary Geology 62: 281-292.

Minell, H., 1978. Glaciological interpretations of boulder trains for the purpose of prospecting in till. Sveriges Geologiska Undersökning C 743. 51 pp.

Nelson, A.R. and D.M. Mickelson, 1977. Landform distribution and genesis in the Langlade and Green Bay glacial lobes, North-Central Wisconsin. Wisconsin Academy of Science. Arts and Letters 65: 41-57.

Nilsson, G., 1973. Nickel prospecting and the discovery of the Mjövattnet mineralization, northern Sweden: a case history of the use of combined techniques in drift-covered glaciated terrain. In: M.J. Jones (ed.). Prospecting in areas of glacial terrain. Institute of Mining and Metallurgy, London: 97-109.

Strömberg, B., 1965. Mappings and geochronological investigations in some moraine areas of south-central Sweden. Geografiska Annaler A 47: 73-82.

Thwaites, F.T., 1943. Pleistocene of part of northeastern Wisconsin. Bulletin of the Geological Society of America 54: 87-144.

Tilas, D., 1740. Tanckar om Malmletande, i anledning af löse grâstenar. Kongl. Svenska Vetenskaps-Akademiens Handlingar 1739-1740:I, 190-193.

Wastenson, L., 1969. Blockstudier i flygbilder. En metodundersökning att kartera markytans blockhalt från flygbilder. Sveriges Geologiska Undersökning C 638. 95 pp.

Whittecar, G.R. and D.M. Mickelson, 1979. Composition, internal structure, and an hypothesis of formation for drumlins, Waukesha County, Wisconsin, U.S.A. Journal of Glaciology 22: 357-371.

# CHAPTER 5

# Glacial flow indicators in air photographs

RAIMO KUJANSUU

*Geological Survey of Finland, Espoo*

## GENERAL

The first indication of a new mineralization is frequently obtained from the ground; an ore-rich erratic unearthed while building a forest road or a geochemical or heavy mineral anomaly revealed during regional mapping. If the case is thought to be worth following up, it is then necessary first to gather a compendium of the existing geological information on the area, one aspect of which will concern the surficial deposits, their origins and the transport distance. This information is required in order to guide the subsequent investigations in the right direction and select appropriate sampling and other methods.

In Finland, where boulder tracing has traditionally formed an integral part of ore prospecting, at least from the beginning of this century (Sauramo 1924), a great deal of work has been done in recent years to gather information together on the transport of boulders and to adapt it into a form in which it will be readily accessible to ore prospectors (Salonen 1987). Since insufficient data are usually available for research at specific sites, or else the data are too general in character, other methods should be available for achieving greater precision. Air photographs and satellite images provide a good starting point for this purpose, since when interpreted by experts they can give a comprehensive picture of the area concerned without necessitating any expensive field investigations, even if no previous mapping data are available for the area.

Satellite data have been used for the general mapping of glacigenic formations over wider areas, and in this way also for describing regional patterns in ice flow directions (e.g. Punkari 1982). Satellite data can be subjected to both visual and numerical interpretation, and the images can be processed in a variety of ways to accentuate certain features, e.g. streamlined accumulation forms. Work has also been done in Finland on the applicability of numerical interpretations of Landsat data to the mapping of Quaternary deposits (Kujansuu and Koho 1982). The intention in this present paper is nevertheless to concentrate on the use which can be made of visual interpretations of traditional air photographs, since it is these that often provide the best and most easily available base map for prospecting purposes.

Visual interpretation of air photographs can yield information on the depths of the Quaternary deposits and the genetic types which they represent, the directions of glacial flow, material transport and the glacial dynamics (e.g. Patterson et al. 1985, Patterson and LeGrow 1986). Black and white photographs to a scale of about 1: 30 000 are usually excellent for this purpose. Colour photographs provide a greater delicacy of detail, but are usually rejected for reasons of cost. Infra-red colour photographs, on the other hand,

Figure 1. Mapping of roches moutonnées near the centre of Helsinki from air photographs. The scale of 1:5 000 allows estimation of the shapes and orientations of individual facets. Published by permission of the National Board of Survey.

can be well worth the extra expense in many cases, since they are able to reveal even quite small bedrock outcrops by virtue of their distinctive colourings (such outcrops being of considerable importance for the next stages in the investigation). A good quality mirror stereoscope such as the Wild Aviopret APT 1 is an invaluable aid when examining these photographs, as its zoom lens will enable direct magnifications of up to 9 × to be obtained. This is of particular significance when using air photographs in slide rather than print form, since details stand out very much better on these. When working in the field one usually has to rely on a small hand lens stereoscope, and many people who use air photographs regularly in this way even learn to view stereo pairs with the naked eye.

Boulder tracing from air photographs can roughly speaking fulfil one of three purposes. 1) Its principal purpose is for the detecting the directions of glacial flow, i.e. the directions of glacial transport, both erratics and till. At the same time it can provide a clear impression of glaciofluvial action, both erosion and deposition. Other geological processes responsible for the movement of material can also be recognized, e.g. flowing water, aeolian action and above all gravitational mass wasting on slopes. 2) It allows preliminary mapping of the nature and thickness of the surficial deposits, essential for planning geochemical sampling. Where necessary, one can also use such photographs to evaluate the genetic types of the formations present and their mode of deposition from the point of view of glacial dynamics, which will provide an impression of the extent of transport, the existence of ice flow stages operating in different directions, any material of a polygenetic nature, etc. 3) Air photographs can also equip one to answer questions with

Figure 2. Glacial grooves (G) and crag and tail forms (C) eroded into a bedrock surface at Aakkenustun-turi in Kittilä, central Finnish Lapland. Aakkenustunturi is a quartzitic erosion relict rising above the peneplain. Stone streams are found on its slopes as a consequence of solifluction (F). The schistosity of the bedrock is indicated by light-coloured streaks (S). Photographed to a scale of 1:30 000. Published by permission of the National Board of Survey.

which the ore prospector is constantly faced, such as whether all the boulders (or components of the anomaly complex) found in the course of the investigation are from the same area or of the same origin. The most important information of all nevertheless concerns the determination of directions of glacial flow.

DETERMINATION OF DIRECTIONS OF GLACIAL MOVEMENT

The indicators of glacial flow observable in air photographs comprise a variety of elements related to the glacial morphology of the area, as discussed by J. Lundqvist in this volume (see also e.g. Prest 1983). Recognition of these from the photographs depends on their size, the scale of the photograph and most of all the skill of the person doing this. For reasons of cost and appropriateness, it is not common to use photographs of such a scale

Figure 3. Streamlined major bedrock landforms (rock drumlins) in the rapakivi granite area of northern Åland showing the glacial flow from north to south. Photographed to a scale of 1:60 000. Published by permission of the National Board of Survey.

that they would reveal striae or other small-scale glacial abrasion forms, but photographs on a scale of 1 : 30 000 are sufficient for the identification of erosional forms with dimensions of approx. 0.5 m in depth, 1 – 5 m in breadth and 10 – 50 m in length on vegetationless bedrock surfaces, or accumulation features of 0.5 – 1 m in height, 1 – 2 m in breath and 10 – 20 m in length.

*Bedrock abrasion forms*

Basins or·fluted surfaces oriented in the direction of ice movement are apt to arise in areas of soft rocks, while roches moutonnées are typical of areas of crystalline schists and igneous rocks. Although individual striae on polished rock surfaces cannot be observed in air photographs, the surfaces themselves are usually sufficiently large that their general orientation can be appreciated (Fig. 1), and their forms normally allow the direction of ice flow to be deduced unambiguously.

It is also possible in some cases for large-scale grooves to develop on bedrock surfaces with depths of 0.5 m or more, breadths of 5 – 20 m and lengths of 20 – 200 m, and also

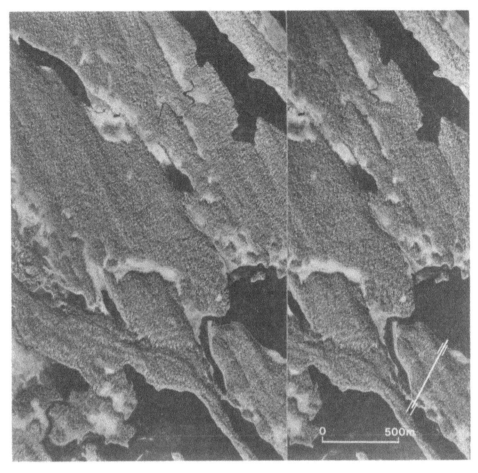

Figure 4. Stereo pair showing drumlins and an esker in Kuusamo, eastern Finland. The major streamlined glacial landforms depict quite clearly the predominant direction of flow in the ice sheet. The eskers mostly run in a consistent direction, but can deviate from this where they follow depressions in the terrain (cf. also Fig. 12). Photographed to a scale of 1:15 000. Published by permission of the National Board of Survey.

crag and tail forms of height 0.5 – 1 m, breadth 1 – 3 m and length 20 – 50 m. These are well visible on air photographs (Fig. 2).

Major bedrock landforms can develop streamlined forms reminiscent of polished rock surfaces (e.g. rock drumlins), if the glacial flow continued for a long period and/or occurred repeatedly with the same orientation, as happened in the case of the Fennoscandian ice sheet in the basin formed by the Gulf of Bothnia and the Baltic Sea proper. Streamlined features of this kind are detectable even on air photographs to a scale of 1: 60 000 (Fig. 3).

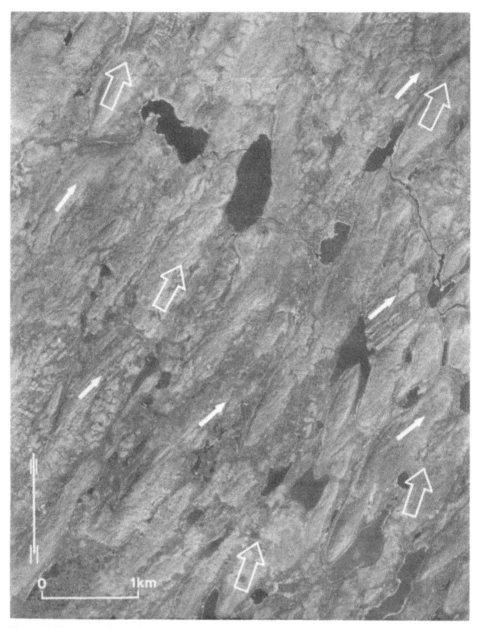

Figure 5. Fluted moraine surfaces and drumlins at Utsjoki, northern Finnish Lapland. The direction of ice flow shifted about 25° further east after deposition of the drumlins, but the flow was so weak by then that it was unable to erode the previously deposited forms away, but instead gouged small grooves in their surface as elsewhere on the low-lying surrounding land and formed also small drumlins and drumlinoids. Photographed to a scale of 1:30 000. Published by permission of the National Board of Survey.

*Moraines*

Landforms composed of glacial deposits, and particularly moraine landforms, are among the most important indicators of ice movement detectable in air photographs. An actively flowing ice sheet will produce streamlined moraine forms, drumlins, which are easily recognizable also in satellite images and on topographic maps when they occur in extensive fields. Drumlins provide very straightforward evidence of the direction of flow of the continental ice sheet (Fig. 4). Although very many types of drumlin exist, the essential point as far as flow directions are concerned is to distinguish those of different ages and belonging to different glacial stages. In Northern Sweden, for instance, two generations of drumlins are found superimposed one upon the other, the younger generation being smaller in size and sharper in outline, although it is the older generation that are more significant in terms of form (see, e.g. Fagerlind 1981, Lagerbäck and Robertsson 1988). A corresponding older stage can be recognized in Northern Finland from its large-sized crag and tail forms, while the younger stage is represented by fluted moraine surfaces. Small-scale fluting can be found on the surfaces of drumlins which differ somewhat in orientation from the drumlin ridges themselves, and these again point to a change in the direction of ice movement following deposition of the drumlins (Fig. 5).

Fluted moraine surfaces have proved to be significant indicators of glacial flow, at least in the area close to the centre of glaciation in Fennoscandia. Although such forms can vary greatly in size, being of height 0.5 – 10 m, breadth 1 – 250 m and length 20 – 3000 m, they are generally sufficiently small or flattened that they cannot easily be recognized from the ground. There would also seem to be some variation in the manner by which they have come about, being attributable either to erosion or to accumulation. The important points here, however, are that they have arisen under the influence of glacial flow, that their longitudinal axis reflects the direction of this flow, and that they stand out from other forms of fluting of bedrock or anthropogenic origin. The large fluting ridges are equivalent to immature drumlins or are drumlinoids and the smaller ones to grooves or small ridges in the till surface (Fig. 6). This frequently seems to be a normal phenomenon on ground moraine surfaces, a fact which conversely may be used to identify ground moraine areas from air photographs (Fig. 7).

Since a glacier will in principle have been flowing towards the retreating ice margin, indirect information on its direction of flow can be obtained from glacigenic elements that indicate the position of this margin, including various moraines of the types referred to as De Geer moraines, annual moraines or washboard moraines (Fig. 8). Certain subglacial hummocky moraines (e.g. Rogen moraines) also tend to run parallel to the ice margin (cf. Lundqvist in this volume), and can similarly be used as indirect indicators of the direction of this margin and thereby of the glacial flow (Fig. 9).

*Glaciofluvial formations*

The mode of creation of a glaciofluvial formation will also depend on the properties of the glacier, including its dynamics. Elongated accumulation forms tend to be arranged either in the direction of glacial flow or at right-angles to this, while erosion forms may reflect either these directions or the gradient of the glacial surface. Eskers, which are particularly easy to recognize and map from air photographs on account of their

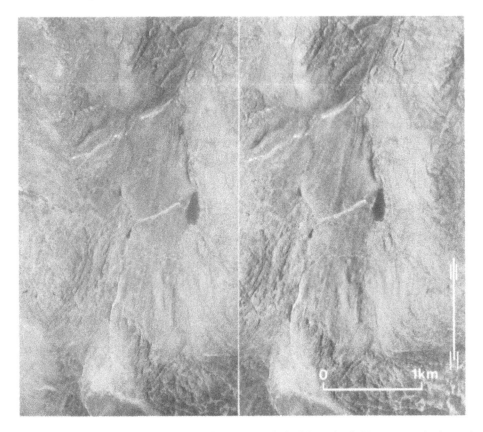

Figure 6. Fluted moraine surfaces at Saariselkä, central Finnish Lapland. The stereo pair shows the deposition of till to have begun on downward sloping land with respect to the direction of glacial flow. The fluted nature of the till cover at the scale used in the air photographs resembles striae on a bedrock surface. Photographed to a scale of 1: 30 000. Published by permission of the National Board of Survey.

characteristic shape, usually provide an excellent impression of the directions of flow prevailing during the deglaciation phase (Fig. 9A), and if any discrepancy arises between these and the moraine landforms this will merely be a sign of a change in flow conditions, e.g. a shift in the centre of glacial flow at the very end of the deglaciation. Eskers usually broadly coincide in orientation with the drumlins and fluted surfaces in the same area.

Ice-marginal deltas and lateral terraces provide data on the direction and course of the ice margin in the same manner as end moraines, and thus these again yield indirect information on the last direction of glacial flow.

The principal indicators of glacial properties among the glaciofluvial erosion forms are various lateral meltwater channels, which can be used for highly detailed construction of the gradient of the glacier surface and the location of the glacier margin in a regional scale (Fig. 10). Various types of overflow channels and marginal or extramarginal channels can also be used to obtain a rough idea of the course of the ice margin in terms of the lateglacial palaeohydrology, and thereby to generate information at a certain level on the directions of ice flow.

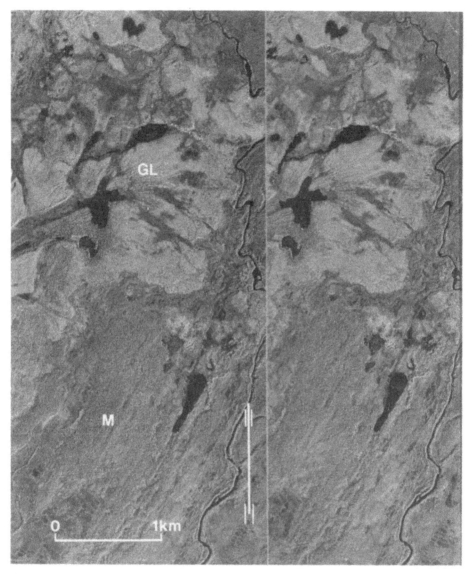

Figure 7. A stereo pair from Enontekiö, western Finnish Lapland, points to pronounced glaciofluvial activity. Meltwater from the glacier has been flowing from left to right in front of the ice margin, depositing its material in an ice dammed lake. The glaciofluvial sediments stand out as light-coloured areas on the air photographs (GL), while the areas of ground moraine are clearly distinguishable as dark-coloured fluted areas (M). The lowest-lying points have a peat cover. Photographed to a scale of 1 : 30 000. Published by permission of the National Board of Survey.

OTHER MATERIAL TRANSPORT PROCESSES

Air photographs can also be used to map other material transport processes, outstanding among which are the various slow or rapid forms of mass wastage on slopes. Boulder enrichment will take place at the surface under periglacial conditions, and these boulders

Figure 8. De Geer moraines at Raippaluoto, among the islands off Vaasa in Gulf of Bothnia, western Finland. These moraines were created parallel to the ice margin, and thus depict a property of the glacier which can be used indirectly to determine the direction of glacial flow. In this example the fluted moraine surface acting as a substrate for the De Geer moraines shows through distinctly. Photographed to a scale of 1 : 30 000. Published by permission of the National Board of Survey.

will then move down the slope to form stone streams or lobes, which are detectable in air photographs (cf. Fig. 2). Till material will tend to slip down a slope in the form of solifluction lobes, often forming a pseudostratigraphy.

Numerous landslides consisting of till are encountered close to the centre of the Fennoscandian glaciation, and in places these have led to major transportation of material and given rise to pseudo-moraine forms (Kujansuu 1975). It is often useful when examining air photographs to gather information on any directions of groundwater flow, especially when investigating a geochemical anomaly or proposing to use geochemical samples in the subsequent stages in the research. Interpretation of the analysis results will be improved by the knowledge of any movement of material in the groundwater.

Air photographs also provide an excellent opportunity to evaluate the role of aeolian activity in material transport, and can be used to identify products of fluvial processes.

CARTOGRAPHICAL PRESENTATION OF DATA

The extent to which the data on the directions of glacial flow and other processes should be collected and the area over which this should be done will depend entirely on how numerous and how obvious these direct or indirect indicators of glacial flow are in the area and how complex the flow model proves to be. One air photograph without stereo

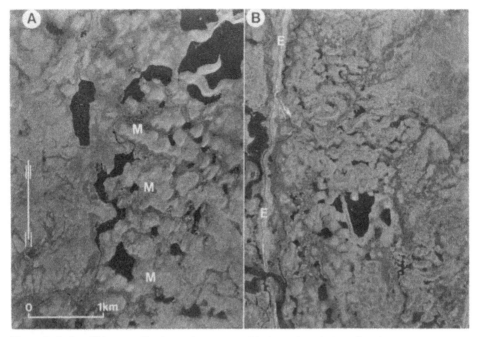

Figure 9. A. Smallish streamlined moraines arranged indeterminately in a direction transverse to the glacial flow at Enontekiö in western Finnish Lapland. Glacial flow was to NNE. B. Subglacial hummocky moraines (M) oriented parallel to the ice margin and at right-angles to the esker (E). The moraines and esker together provide indirect evidence of a S - N glacial flow. Photographed to a scale of 1 : 30 000. Published by permission of the National Board of Survey.

viewing or any more precise analysis of the landforms may well be sufficient to give a clear impression of the direction of glacial flow in an area with a typical drumlin terrain, whereas the gathering of adequate information in an area of inactive ice may require a survey to be made of air photographs covering hundreds or thousands of square kilometres. In the latter case at least, it is necessary to be able to display the results in cartographical form. In terms of time, this still will not require more than a fraction of the input called for if one attempts to survey and map the area in the field.

The transfer of air photograph data to a base map can be accomplished most easily using a sketchmaster device such as the Zeiss Aerotopo LUZ.

PHOTOGEOLOGICAL RECONNAISSANCE MAPPING

Data on the surficial deposits in an area are required in cartographical form when it is intended to commence geochemical sampling, which when conditions are difficult is often best undertaken in winter, travelling by snowmobile. This means, however, that no very reliable picture of the nature of the material can be gained at the time of sampling because of the snow cover unless one has access to a sampling grid superimposed on a map of surficial deposits. Similarly the planning of sampling transects on a map in advance will ensure the appropriateness of the sites chosen and reduce unnecessary work.

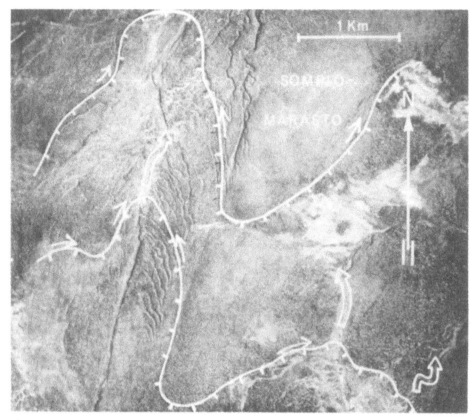

Figure 10. Lateral and proglacial meltwater channels (arrows) at Kittilä in central Finnish Lapland which allow detailed reconstruction of the course of the melting ice margin (white line). The lateglacial palaeohydrology provides information on the properties of the ice sheet that allows indirect conclusions to be reach regarding the NNE directions of flow. Photographed to a scale of 1 : 30 000. Published by permission of the National Board of Survey.

It is also essential when interpreting the results of the analyses to know what kind of material each sample was taken from.

The classification of deposits to be used depends on the purpose for which the map is to be constructed and the nature of the geology of the area concerned. The most important thing is to distinguish the genetically distinct units, especially those differing in their transport history. Moraines, for example, can be divided into two major classes, those deposited by active and passive ice, in addition to which it is useful to recognize thin moraines, or areas with a thin cover of loose deposits in general (< 1 m), as a separate class. Other significant categories are glaciofluvial, fluvial, aeolian, lacustrine, marine, littoral and organic deposits. Fine water-lain sediments and peat layers, which together cover 25% of the land area of Finland, for instance, to a depth of at least 1 m, constitute a hindrance to indicator tracing.

The known directions of material transport and observations of directions of ground-water flow should also be marked on the map. Ore prospectors will gain important

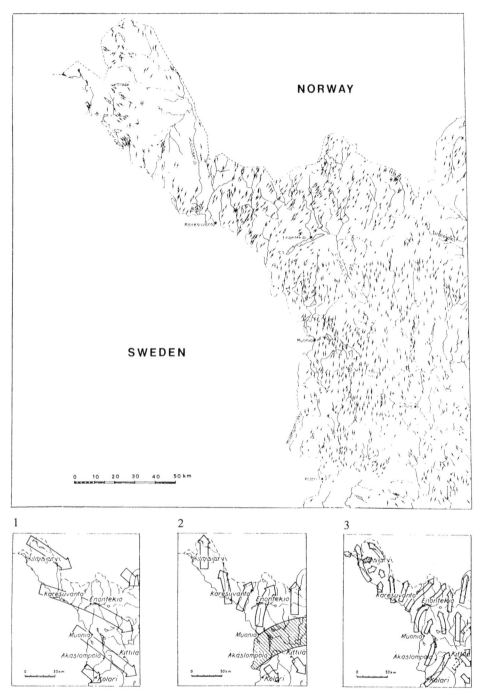

Figure 11. Fluted moraine surfaces in western Finnish Lapland and flow stage maps based on these (Kujansuu 1967). The large-scale crag and tail forms represent the oldest flow, that associated with the early stage in the glaciation (map 1), the streamlined moraine forms and eskers reflect the main flow stage (map 2) and the fluted moraine surfaces and lateglacial palaeohydrology can be used to reconstruct the last flow direction, corresponding to the deglaciation stage (map 3).

information from notes on the bedrock, especially outcrops, and these should be marked on the map, both those directly observable from the air photographs and those deducible from their topographical position, their form (geomorphology), or the steepness of the slope, or which have been exposed from beneath the surficial deposits by processes of glaciofluvial or fluvial erosion.

The thickness of the surficial deposits is also an important consideration in ore prospecting, being relevant to geochemical sampling and to the choice of appropriate methods for exposing the bedrock where necessary or obtaining a bedrock sample by one means or another. This can be estimated from the type of formation involved. Eskers and various moraine landforms are usually of a considerable thickness, as are valley fill deposits. The topography in fact leads to a general division of the surficial deposits into thin layers on the hills and thicker layers in the valleys. A thin cover of surficial deposits will usually reveal faults and other structural features and various plastic deformation products in the bedrock, and sometimes even variations in rock types.

An experienced research worker can draw a reasonably accurate photogeological map of an area of about 100 km² in a day from air photographs of scale 1: 30 000. Relatively little time is taken up in transferring the patterns to a base map using a sketchmaster.

CASE STUDIES

### Determination of glacial flow directions in western Finnish Lapland from air photographs

Directions of flow of the continental ice sheet in western Finnish Lapland were deter-mined from air photographs during the early 1960's in connection with the general mapping of Quaternary deposits (Kujansuu 1967). The principal body of data comprised fluted moraine surfaces of various kinds (many of the illustrations to this paper are from western Lapland), but use was also made of the lateglacial palaeohydrography and other data on surficial deposits extracted from the air photographs. The majority are related to glacial flow during the last stage of the deglaciation, but some signs of older, deviant directions of flow were also recognized. The data were assembled using a sketchmaster into a map which provides a general picture of the distribution of flow directions (Fig. 11), which in combination with other observations made in connection with mapping and additional topographical information could be used to construct maps of the various ice flow stages. The results of detailed fieldwork carried out more recently have been used to fill in and in places adjust this picture, especially as far as the chronology of the flow stages is concerned, but did not yield anything essentially new with regard to the directions prevailing during the last flow stage.

### Directions of glacial flow in the Riihimäki map sheet area in Southern Finland

In addition to a number of eskers and the major Salpausselkä I end moraine, large numbers of De Geer moraines are to be found in the Riihimäki area, which are easy to identify in air photographs. The radially oriented eskers and the end moraine and De Geer moraines running parallel to the ice margin enable an ice flow pattern to be reconstructed which provides a clear picture of flow conditions during the deglaciation phase (Fig. 12).

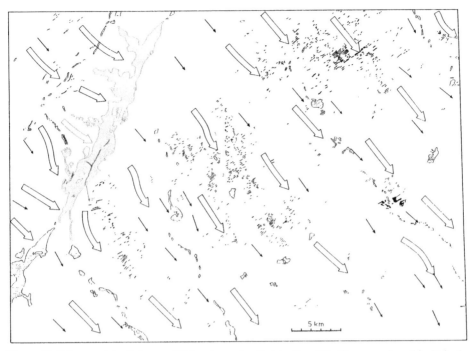

Figure 12. Eskers and the Salpausselkä I end moraine (hatched) and De Geer moraines (black) in the Riihimäki map sheet area (Tynni 1969: Fig. 3 and appendix 1). The orientations of these glacial landforms are used to construct an ice flow pattern (open arrows), which gains support from striae observations (small arrows). This map serves to demonstrate the sensitivity with which landforms of different kinds reflect glacial flow directions and their changes.

The glacial flow which preceded the Salpausselkä stage was from a approx. 15 – 20° more northerly direction than that prevailing during the Salpausselkä I stage itself. Clearer evidence for this is obtainable from the interpretation of the glacial landforms than from striae observations made during mapping in the field (cf. Tynni 1969).

SUMMARY

In order to be successful, indicator tracing requires data on surficial deposits, and especially on material transport in these. This paper sets out from air photograph data, which are capable of supplying a broad spectrum of information on the surficial geology, topography and vegetation of an area. Various erosional and depositional landforms attributable to glacial flow can provide direct or indirect information on the directions of transport within the ice, especially the last such direction, and this can be filled out with observations on glaciofluvial activity, which are capable of contributing further data on features connected with the form and dynamic properties of the glacier. More far-reaching assessment of the nature of the surficial deposits will allow appropriate planning of the sampling stage in the research and assist in the interpretation of the analytical results. Air photography is a rapid and relatively inexpensive method which should automatically be included as a preliminary research stage in any indicator tracing project.

## REFERENCES

Batterson, M. and P. LeGrow, 1986. Quaternary exploration and surficial mapping in the Letitia Lake area, Labrador. Current Research (1986) Newfoundland and Labrador Department of Mines and Energy, Mineral Development Divison, Report 86-1: 257-265.

Batterson, M., D.Taylor and S.Vatcher, 1985. Quaternary mapping and drift exploration in the Strange Lake area, Labrador. Current Research (1985) Newfoundland and Labrador Department of Mines and Energy, Mineral Development Division, Report 85-1: 4-10.

Fagerlind, T., 1981. Glacial development in the Pajala district of northern Sweden. Sveriges Geologiska Undersökning. Ser. Ba 27, 118 pp.

Kujansuu, R., 1967. On the deglaciation in western Finnish Lapland. Bulletin de la Commission géologique de Finlande 232, 98 pp.

Kujansuu, R., 1975. On landslides in Finnish Lapland. Geological Survey of Finland Bulletin 256, 22 pp.

Kujansuu, R. and S. Koho, 1982. On the suitability of the Landsat data for the general geological mapping of Quaternary deposits in northern Lapland. The Photogrammetric Journal of Finland 9(1): 65-75.

Lagerbäck, R. and A.-M. Robertsson, 1988. Kettle holes - stratigraphical archives for Weichselian geology and palaeoenvironment in northernmost Sweden. Boreas 17: 439–468.

Prest, V.K., 1983. Canadas heritage of glacial features. Miscellaneous Report 28. Geological Survey of Canada.

Punkari, M., 1982. Glacial geomorphology and dynamics in the eastern parts of the Baltic shield interpreted using Landsat imagery. The Photogrammetric Journal of Finland 9: 77-93.

Salonen, V.-P., 1987. Observations on boulder transport in Finland. In: R. Kujansuu and M. Saarnisto (eds.), INQUA Till Symposium, Finland 1985. Geological Survey of Finland, Special Paper 3: 103-110.

Sauramo, M., 1924. Tracing of glacial boulders and its application in prospecting. Bulletin de la Commission géologique de Finlande 67, 37 pp.

Tynni, R., 1969. Explanatory text to the map of Quaternary deposits. Geological Map of Finland 1:100 000. Sheet 2044, Riihimäki. 95 pp.

# CHAPTER 6

# Boulder transport in shield areas

MICHEL A. BOUCHARD
*Department of Geology, University of Montreal, Québec, Canada*

VELI-PEKKA SALONEN
*Geological Survey of Finland, Espoo*

## INTRODUCTION

Known transport distances of glacial boulders appear to range from short, on scales of hundreds of meters to few kilometers, leading to the occurrence of so-called 'local' boulders, to very long, on scales of hundreds of kilometers, producing so-called 'far-travelled erratics'. Long distance transport may exceed 1800 km (Prest and Nielsen 1987) for the North American Ice sheet complexes.

Within the glaciated shield areas of Québec and Finland, boulders occur often at the surface of glacial deposits as 'boulder fields', 'boulder mantles', or they are concentrated in topographic lows as 'boulder streams'. Most of these boulders have a seemingly short glacial transport history and consequently belong to the 'local' variety of boulders. Salonen (1986, 1987) found a median length of about 4 km for the distance of dispersal of surface boulders in Finland with few fans exceeding 100 km in length. The occurrence of bouldery morainic surfaces, the local character of the surface boulders, together with the abundance of bouldery sandy till, a feature derived from the predominantly granitic subcrop, all appear to characterize the central areas of glaciation of the Northern Hemisphere.

The purpose of this paper is to present an overview of the glacial geology of boulders over the glaciated shields based on our individual and common experience in the Québec part of the Canadian Shield, and in the Finnish part of the Fennoscandian Shield (Fig.1). In this paper, we discuss the occurrence, the origin and the dispersal of glacial boulders with emphasis on their usefulness to understand the genesis of till, to trace back ore boulders to their sources, and to decipher complexly changing ice-sheet configurations.

Some importance is given to the study of the relationship between surface boulders, glacial morphology, and till lithofacies (Bouchard et al. 1984) and on the influence of the glacial dynamics on the distances of boulder transport, as established by the method of Salonen (1986).

## GLACIAL BOULDERS

Boulders can be defined as detached rock fragments, sedimentary detrital grains, with a diameter exceeding 256 mm on the Wentworth (1922) scale. Glacial boulders are rock fragments that have been transported over variable distances by glaciers or ice sheets; it is generally understood that the glacial boulders originate mainly as a result of glacial

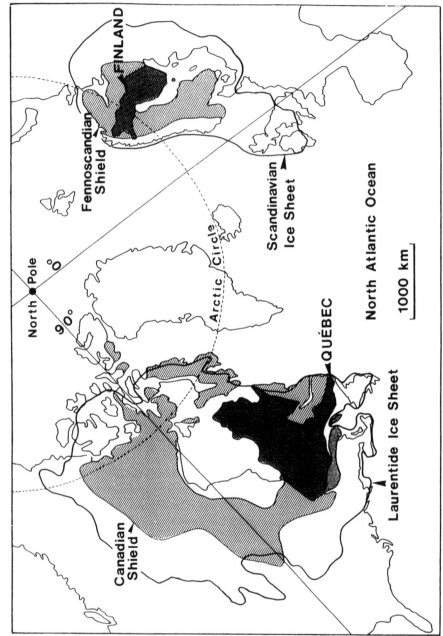

Figure 1. Location map. The discussed areas are situated on Precambrian Shields, centrally in relation with the Late Wisconsin/Weichselian Laurentide/Scandinavian Ice Sheets.

erosional processes such as quarrying and plucking. They are called erratics, when deposited at some distance from the outcrop from which they were derived. The source rock is their provenance. A glacial boulder is referred to an indicator, if its provenance is known, in other words, if it can be traced back to a relatively well defined bedrock source on the basis of its lithology. Indicators are used to decipher boulder transport, the study of which includes the determination of the glacial ice flow directions and the transport distance; the latter in turn is used to infer the persistence or duration in time of the flow events and finally the style or mode of transportation and deposition by the flowing ice.

Boulder studies have had two significant applications. First, the use of indicator boulders and the study of boulder transport have found direct applications in the determination of the provenance of associated mineralized fragments. Boulder tracing has been of great importance in ore exploration both in Scandinavia (Sauramo 1924, Grip 1953, Hyvärinen et al. 1973, Saltikoff 1984) and in Canada (Dreimanis 1956, Shilts 1976, DiLabio 1981). Boulder tracing involves the recognition and reconstruction of dispersal fans or trains leading up-glacier to the source area, as simply as the smoke plume leads back to the chimney. When the matters are simple, this method is probably the most straightforward mean of mineral exploration known to man. However, matters are frequently not as simple, and it often becomes necessary to be able to determine the nature and origin of the boulder population in order to connect the head of a dispersal train to a precise location in the subcrop, let alone the need to resort to some knowledge of the systematics of boulder transport in order to even reconstruct a boulder train. In that case, closer analysis of the associated glacigenic deposits is required to support boulder tracing.

An additional outcome of boulder transport studies has been the reconstruction of the first-order features of the regional glacial ice flow patterns with corollary informations on the changing ice sheet geometry and the successive positions of ice flow centers within given glaciated areas. This involves the reconstruction of the glacial flow lines at various scales.

Boulton et al. (1985) have pointed out the importance of particle trajectories in space and time as characteristics of ice sheets. According to them, the mapping of maximum rock and mineral dispersal pathways is a technique by which ice sheet models can be tested against empirical data. Based on regional dispersal data, including boulders, Shilts and his co-workers (1979, 1985) have attempted to reconstruct flow lines and the configuration of the ice dispersal centers for the central parts of the Laurentide Ice Sheet west of Hudson Bay. Bouchard and Marcotte (1986) and Klassen and Thompson (1987) used the same type of evidence for the reconstruction of the Laurentide Ice sheet east of Hudson Bay. Dyke et al. (1987) have studied long distance boulder dispersal from known sources to set some contingency on the reconstruction of the North American ice sheet complex during the Late Wisconsinan.

TRANSPORT DISTANCES

The glacial transport of debris, eventually to be deposited as till, is a complex system which involves the conditions of flow of the ice and the conditions of debris acquisition and deposition by the glacier. Boulders represent a sample only of the whole of the transported glacial debris by a given mass unit of glacier ice. Therefore, the transport

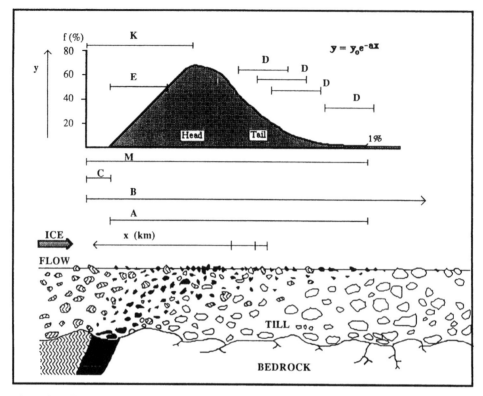

Figure 2. Different methods to determine the transport distance of glacially derived particles using a distinct indicator rock source (colored with black). A = The length of a boulder fan, B = The distance from proximal contact to farthest travelled particle, C = The reaching-surface distance, M = The maximum distance, D = The half-distance, K = The distance between the proximal contact and maximum frequency, E = The renewal distance, x = distance from observed particle to its source. For further explanations, see Table 1.

distance of boulders can only reflect part of the whole sedimentary system. As defined above, the length of boulder transport is the distance between a point where a boulder is observed in the glacial drift and its provenance in the bedrock. Although this appears quite simple, there are a number of different ways to measure the transport distance. These are reviewed first. In addition, many factors determine the transport distance of boulders. These are discussed afterwards.

*Measurement and significance*

The distance at which boulders are transported by glacier ice can be measured in a number of different ways (Fig. 2), based on the common observation that glaciers disperse material so that a given rock type reaches a maximum frequency in the till close to its source, followed by an exponential decrease in the direction of glacial transport. The basic observation of Krumbein (1937) has since been tested in several studies of

Table 1. Summary of the various parameters used to characterize the transport distance and the main factors believed to determine the specific parameters.

| Para-meter | Definition | Reference | Main factor |
|---|---|---|---|
| A | Length of boulder train | Sauramo 1924, and many others | The mode of glacial deposition |
| B | Distance between the proximal contact and the farthest travelled particle derived from the provenance | Prest & Nielsen 1987 | Distinctiveness, relative hardness |
| C | Distance at which the first indicator appears at the surface | Drake 1983 | Thickness of till; rate of erosion vs. rate of deposition |
| X | Distance to the provenance area | Many | |
| K | Distance between the major concentration of surface clasts and the bedrock ore zone | Lee 1965, Shilts 1973, Salminen 1980 | Defines the head of the dispersal area |
| M | Maximum distance, at which the frequency passes below 1% | Bouchard et al. 1984 | Rock resistance |
| E | Renewal distance; the distance in which the frequency increases from 0 to 50% | Peltoniemi 1985 | Relative erodibility |
| D | Half-distance, the distance in which the frequency falls to half of its original | Krumbein 1937, Gillberg 1965, Perttunen 1977 | Size of outcrop |
| -a | Rate of decrease of frequency | | |
| GM | Geometric mean of transport distance distribution (Figure 3) | Salonen 1986 | The mode of glacial deposition |
| SD | Standard deviation (Figure 3) | Salonen 1986 | The mixing rate |

glacial dispersal of rocks and minerals, (Gillberg 1965, Marcussen 1973, Shilts 1976, Perttunen 1977, Bouchard et al. 1984, Peltoniemi 1985, Salonen 1986) and it is still a suitable model which does apply to the boulder and to finer grain-size fractions of the till.

The parameters most frequently used as an index of transport distance are distances B , D and K (Fig. 2) and the main factors controlling these various parameters are summarized in Table 1.

Based on the same exponential decrease model, a different method has been newly suggested by Salonen (1986) and successfully applied by him to most of the dispersal data in Finland. In this 'transport distance distribution' (TDD) method, a sample of a number of boulders of different lithologies is treated instead of a single indicator rock type. The true distance to the provenance area measured for every lithologic type upstream along the glacial flow vector is plotted in an orderly cumulative curve (Fig. 3). Plotted on a lognormal scale, the distribution can usually be fitted with a straight line that allows the graphic estimate of such transport distance statistics as geometric mean (GM) of the distance to provenance for a given boulder population and its standard deviation (SD). The lognormality of the best fitting curve can be tested by using chii-square test (Sinclair 1983).

In central Québec, Bouchard et al.(1984), using parameters K and M, observed that the

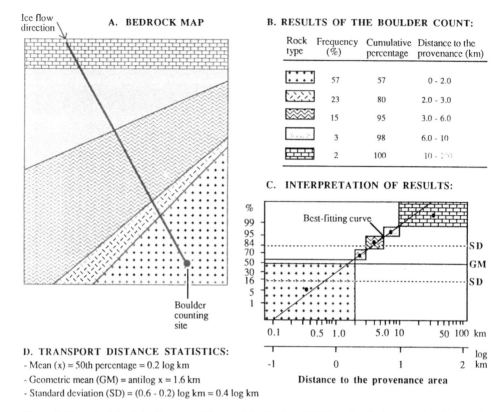

D. TRANSPORT DISTANCE STATISTICS:
- Mean (x) = 50th percentage = 0.2 log km
- Geometric mean (GM) = antilog x = 1.6 km
- Standard deviation (SD) = (0.6 - 0.2) log km = 0.4 log km

Figure 3. The principle of the Transport Distance Distribution (TDD) method in determining the length of glacial transportation (Salonen 1986). A. The lithology for every particle (boulder) at sampling site is determined. B. The distance to the provenance area is measured for every rock type in sample. The cumulative frequencies of rock classes are put in order of increasing remoteness of its source. C, D. Transport distance statistics are determined on probability paper.

characteristics of the transport distance of the coarse fraction of the till, including boulders, varies according to the depositional setting as indicated by the morainic landforms and the till facies within which the boulders are associated or from which they are derived. They argued that understanding the anatomy of the till sheet was the key to explain the glacial dispersal patterns and to trace back ore boulders to their sources.

In Finland, Salonen (1986), using parameters D, GM and SD, observed that the characteristics of the transport distance of boulders varies systematically within a given area according to the depositional setting, as established by Bouchard and others (1984), and from area to area, according to glacier flow conditions and style of deglaciation.

*Factors affecting transport distance*

Factors affecting the distance of glacial transport of boulders are geological, lithological, and glaciological in nature.

A commonly cited geological factor is the size of the source area which, according to Peltoniemi (1985) is the main factor affecting the transport distance of the coarse fraction of till (Table 1). Other geological factors are in some way objective, preset conditions, under which boulders are produced and transported. For instance, shield areas are particularly proned to yield angular boulder-size clasts because of the abundance of structural discontinuities, mostly joints, in the Precambrian, granitoid rocks as well as in the intervening greenstone belts. Therefore, the abundance of boulders in the glacial drift of the central areas of glaciation can be viewed either as a sign of intense and late glacial erosion, short glacial transport, or more simply as a reflection of the propension by the eroded bedrock to produce boulders (Shilts et al. 1987).

The most important single factor affecting the transport distance of boulders is lithologic in nature, namely the resistance of different lithologies to abrasion and crushing during glacial transport. Once detached from their source, newly formed subglacial boulders become sedimentary grains which may be abraded, rounded, faceted, or crushed to finer sizes during transportation, a process generally referred to as comminution (Elson 1961).

The comminution results in a decrease in size of individual particles and in an increase in the total number of particles (Drewry 1986). Consequently, the travel distance of a given boulder can only merely be the distance over which it passed from a single fragment well over 256 mm in size to a number of smaller fragments, including silt and clay, by virtue of the fact that it was crushed or abraded during transportation.

Rates of comminution are closely related to the preset geological conditions. For example, as noted by Gillberg (1965), a dilution by softer rocks, such as limestone and shale, in till will result in a lower rate of communition for more resistant rocks, e.g. sandstones, than for the same rocks within a quartzitic matrix in a till derived from granitic terrains. Comminution is most effective in the basal transport where grain to grain contacts are frequent (Kinnunen 1979). Goldthwait (1971) has estimated, that less than 0.1% of all the fragments from a any given source lithology remains polymineralic beyond 35 km (see also Dreimanis and Vagners 1971).

Haldorsen (1981) has shown the profound effect that comminution has on the composition of the various size fractions of till. For these reasons, the coarse gravel-size fraction of till may be biased in composition toward lithologies of the bedrock sources which are more proned to quarrying and the fragments of which are more resistant to comminution. A corollary is that a high proportion of boulders in till may be an indication of its 'immaturity' and short glacial transport.

Glaciological factors are numerous as they range from the various factors that control the subglacial erosion, the mode of transportation and the rate of flow, and finally the sedimentation of the transported debris. Perhaps the most determinant single factor is the mode of transportation, i.e. englacial versus subglacial transport, since this largely determines the amount of grain to grain and grain to bed contacts of the transported debris.

Within the subglacial mode of transportation, all preset geological conditions and the lithological factors being equal, the transport distance of boulders can be considered as resulting from the distance that separates areas or 'zones' of the glacier bed where erosion is the dominant process from other areas or 'zones' downglacier where deposition is the dominant process. The number of times that bouders may have gone through such 'cycle' of being supplied to, moved with, and deposited by the glacier ice forms another control. This in turn appears to be related mostly to the mode of deglaciation and concommittant glacial activity (Salonen 1986).

SURFACE BOULDERS

Surface boulders are either derived from the underlying till or they represent part of a distinct deposits. Consequently, boulder transport may or may not reflect the dispersal characteristics of the underlying sediment.

Dense bouldery surfaces of the central areas of glaciation as well as other occurrences of surface boulders overlying till in more peripheral areas are generally thought of as lag concentrates produced by washing, deflation, or frost-heaving of till surfaces. The density of the boulder field reflects the original abundance of boulders in the till and the amount of surface lowering since the primary deposition. Lag origin for most of the 'boulder streams' in glaciated areas can be inferred from the fact that they occupy meltwater channels (Bouchard 1980). Otherwise, surface boulders may not be the result of some special concentration process, but may merely represent the original distribution of boulders in the glacial deposits that outcrops at that particular location. In those situations, the composition of the surface boulders reflects the dispersal characteristics of the underlying till.

Alternatively, surface boulders within fields or mantles, can be viewed as part of a single deposits, a 'bouldery' unit within a till sheet. There are at least two different possible origins for such a distinct unit, namely englacial or subglacial. The englacial origin implies that the deposition occurs as the letting down of englacially transported boulders, concentrated during the thinning and stagnating of the ice (Shilts 1973). The deposits should then be referred to as an 'ablation mantle', and should be considered as a particular facies of ablation deposits.

The concentration of boulders may also have occurred at the time of deposition in supraglacial marginal environments (Lawson 1981), as bouldery debris-flows or washed supraglacial melt-out tills. The sediment could in fact be viewed as some syndepositional lag deposits. The subglacial origin implies that the boulders are a particular facies of subglacial till, an immature till layer (Bouchard et al. 1984) made out of coarse, slightly comminuted, locally derived boulders ('till at birth', Minell 1980). In those situations, the composition of the surface boulders do not reflect the dispersal characteristics of the underlying till.

Close to the marginal areas of the continental glaciers the composition of the surface boulders is generally consistant with an origin as en- or supraglacial debris with a long and complex glacial transport history. The depositional setting is commonly indicative of a development as ablation or lag derived from marginal deposit. In contrast, the composition, as well as other attributes of the boulders making out fields in the central areas of glaciation is generally consistent with origins ranging from englacial to sub-glacial, as discussed in the following section.

DEPOSITIONAL SETTING

Surface boulders are associated with various types of morainic terrains, the assemblage of which characterize the glacial geology of central shield areas. The transport distance of boulders over a given morainic terrains differs consistently from that of surface boulders occuring over other morainic terrains. It is therefore concluded that the transport distance must be related to the overall sedimentology of the till sheet and to the associated development of the morainic terrains. Facies modelling is applied here for the purpose of grouping the depositional environments and describing the transport paths of boulders in those different systems.

The anatomy of till sheet over central Québec is sketched in Figure 4. Four distinct lithofacies of till, based on differences in their textural, structural and compositional attributes, are recognized. While they all occur in a definite vertical sequence, various assemblages of lithofacies underlie distinct types of morainic terrains, namely drumlinized and fluted ground moraine, Rogen (ribbed) moraines and associated low relief fluted hummocky moraines, and unpatterned unfluted hummocky moraines.

The various attributes of the till facies are summarized as Table 2. They are described and discussed in detail by Bouchard (1980,1986) and Bouchard et al. (1984) .

Lithofacies 1 is a massive, matrix supported fine-grained diamicton of predominantly local provenance. It occurs mostly on interfluvial areas where it underlies drumlinized and fluted ground moraine (Fig. 4). The study of the coarse-fraction of this till facies have led to the following observations: a. the deposit systematically contains upward an increasing proportion of distantly derived fragments, b. down ice from a given source, the abundance of the fragments of that source peaks within 2 km from which point it follows the exponential decrease rate discussed earlier, c. transport distance, defined as M (Fig. 2) ranges from hundreds of meters to tens of kilometers, the maximum observed being 15 km. The upper surface of lithofacies 1 was eroded or moulded, syn- or post-depositionally, into long drumlinoid features, and, at places, to prominent fluted surfaces (Bouchard 1980).

Lithofacies 2 is a massive to stratified matrix supported fine-grained diamicton of variably local to distal provenance (Fig. 4). In contrast to facies 1, facies 2 occurs most commonly in topographic lows where it forms the bulk of the till sheet. It is slightly better sorted while containing a similar proportion of silt and clay; above all, it shows a distinct layering of the type commonly observed in a variety of diamicton described as Kalix till in Sweden (Lundqvist 1969).

This facies, described in detail by Bouchard (1980,1986), has been interpreted by him as a basal melt-out till, the layering being inherited from the original layered disposition of the debris during its near-base englacial transport. Again, compared to facies 1, the coarse fraction of facies 2 is generally composed of a higher proportion of distantly derived clasts, probably reflecting the less intense comminution associated with the englacial transport of the debris. Exceptionally, in areas where facies 2 occur in the lee of a predominant rock source, the abundance of fragments from that provenance is overwhelming, and facies 2 then appears to be of more local in lithological origin than the stratigraphically lower facies 1 in adjacent terrains.

The detailed study of the coarse-fraction of this till facies have led to the following observations: a. similarly to the facies 1, the deposit contains systematically an upward increasing proportion of distantly derived fragments and, b. down ice from a given source, the abundance of the fragments of that source peaks within 4 km.

Lithofacies 3 is a massive, matrix supported, coarse-grained diamicton, of almost exclusively local derivation. Facies 3 has not been observed elsewhere than in topographic lows, associated with, and overlying facies 2 (Fig. 4). An erosional contact separates the two facies and lenses of the underlying facies 2 can be observed, deformed and folded, within bouldery facies 3. The boulders which form the bulk of facies 3 are derived from within few km and, apart from the erratic material provided by the inclusion of the underlying unit, is generally almost devoid of foreign bedrock material.

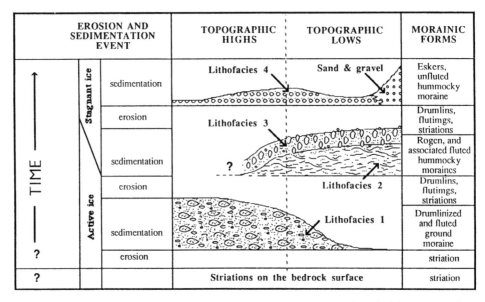

Figure 4. Landforms and till lithofacies in central Québec. The successive glacial events produce various morainic landsforms depending on topography of glacial bed. The lithofacies have been indicated with numbers 1–4.

Altogether, facies 2 and 3 are the sediments which underlie Rogen (ribbed) moraine, and presumably, the associated fluted hummocky moraines (Fig. 5). As shown by Bouchard (1986), the bulk of Rogen moraines is made out of overridden thrusted slabs of facies 2. The upper erosional, unconformable contact of facies 2 forms the fluted and lineated surface of Rogen moraines and adjoining or associated hummocky terrains. Facies 3 forms the bouldery surfaces, the boulder fields, trains and fans, commonly seen overlying these terrains. As noted by many, boulders in these fields are frequently lodged into the surface of the moraine; flutes extend down glacier from some of them. All these features suggest the subglacial origin of facies 3 and the expression 'immature till' used by Bouchard et al. (1984) is based on such an interpretation.

In different and widely spaced areas of Québec, the transport distance of the boulder fields overlying Rogen moraines have been observed to be in the order of 5 km (Bouchard 1980, 1986, Bouchard and Marcotte 1986, Richard et al. 1981). This suggests that Rogen moraines may be formed at least 5 km behind the glacier margin.

Lithofacies 4 is strongly variable in texture and structure, ranging from interstratified sand, gravel, and massive sandy diamicton, to a bouldery diamicton of predominantly distal provenance. This facies is commonly observed in areas of unfluted hummocky moraine indicative of final deposition by stagnant ice. Stratigraphically it overlies all previous facies both in interfluvial areas and topographic lows (Fig. 4). Some of the chaotic landscape associated with Rogen moraine terrains is believed to result from the unconformable deposition of facies 4, in thick piles and hills, above the Rogen landform.

The detailed study on the composition of the coarse fraction of facies 4 shows that it

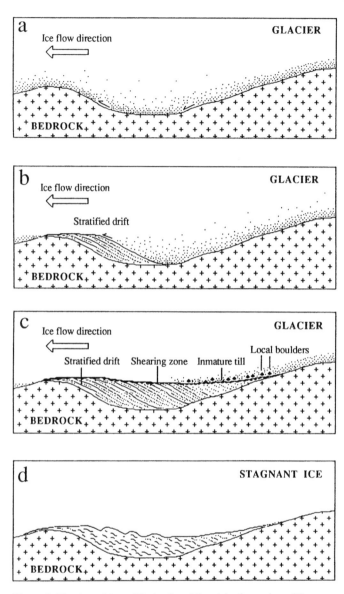

Figure 5. The deposition of facies 2 and 3 and the formation of Rogen moraines (Bouchard 1986).

contains a significantly higher proportion of more distantly derived fragments than any of the underlying facies in a given location. This suggest that facies 4 must be derived partly from the same debris that eventually forms the first 3 facies, augmented by distinct englacial debris.

The development of the assemblages of facies and landforms of Figure 4 has been schematically depicted in Figure 6. The deposition of facies 1 occurs first, well behind the glacier margin. Facies 2 is thought to be associated with a net freeze-on (regelation) zone

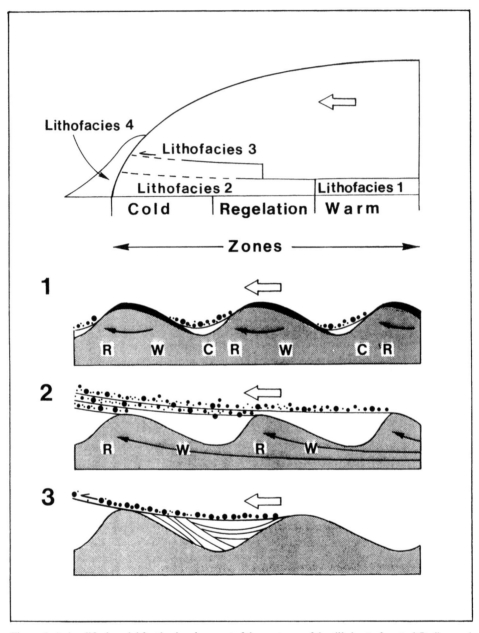

Figure 6. A simplified model for the development of the anatomy of the till sheet of central Québec and for the formation of the various morainic landforms. The open arrow shows the direction of glacial flow. The temperature zones refer to basal temperatures less than (cold) or equal (warm) to the pressure melting point (Boulton 1972). The schemas 1, 2, and 3 illustrate the origin for respective lithofacies 1, 2, and 3, and their relation with bedrock topography. Black arrows indicate the passage of water.

where the layered structure of the deposit is developed. It may be deposited concurrently, sideway or down ice from the point of deposition of facies 1, according to the configuration of the thermal mosaic of the glacier bed at any one time (Hughes 1981). While facies 2 is being deposited, the upper surface of facies 1 may be moulded into a drumlinized and fluted ground moraine.

Facies 3, as suggested on Figure 5, is developed some 5 km behind the ice margin, following the deposition of facies 2 in the topographic lows. While facies 3 is being deposited, the upper surface of facies 2 is becoming fluted and partially drumlinized, leading to the surface lineation of Rogen moraines. In the same time the upper surface of facies 1 will at places be remoulded into smaller superimposed drumlins (Bouchard 1986). The development of facies 3 represent the result of a late phase of quarrying during deglaciation.

Facies 4 is deposited in the marginal environment of a glacier. When this facies is being deposited, the surface of all previous deposits is being cut, trenched, and washed at places, into meltwater channels and local lag deposits.

### TRANSPORT DISTANCE AND MORAINIC TERRAINS

According to the observations of Bouchard et al. (1984), the various facies of till can be arranged in order of decreasing proximality from facies 3, 1, 2 and 4. Consequently, surface bouders derived from or associated with these various facies and landforms should exhibit similar transport distance characteristics.

A systematic study of the transport distance of surface boulders in Finland has shown a consistent variation in the dispersal characteristics associated with various depositional settings as shown by various glacial landforms (Fig. 7). The results are in line with the observations from central Québec.

For 'active ice hummocky moraines' the geometric mean varies between 0.4 and 3.0 km (Fig. 7A). Active ice hummocky moraines as used here include morainic terrains described elsewhere as Rogen (ribbed) moraines and fluted hummocky moraines. The surface bouders associated with these landforms were shown to be facies 3 with the greatest proximality, with M in the order of 5 km (Table 2).

The geometric mean for the transport distance in 'cover moraine' areas (Fig. 7 B) varies between 0.8–10.0 km, while the surface boulders of drumlins have a geometric mean of 5–17 km (Fig. 7 C) as a parameter for their transport distance distribution. In the model of Bouchard et al. (1984) both lithofacies 1 and 2 have features connecting it with this population. There is no direct equivalent for the term 'cover moraine' in central Québec; it can be considered unfluted and undrumlinized ground moraine. Part of such terrains can likely be composed of facies 1 and 2. Drumlins are probably developed mainly from lithofacies 1 and the results shown as Figure 7 C are in agreement with the characteristics of that facies in central Québec, with M around 15 km (Table 2).

Mixed transport population (Fig. 7 D) has been encountered in ground moraine, especially in northern Finland. The transport distance (GM) of surface boulders varies within a wide range (1.0–25 km). A variable proportion (approx. 1–50%) of boulders represents the mixing of materials of earlier glacial cycles. It may have had a long and complicated transport history which is difficult to trace out. This kind of multistage transport is found especially in the central areas of continental ice sheets where the

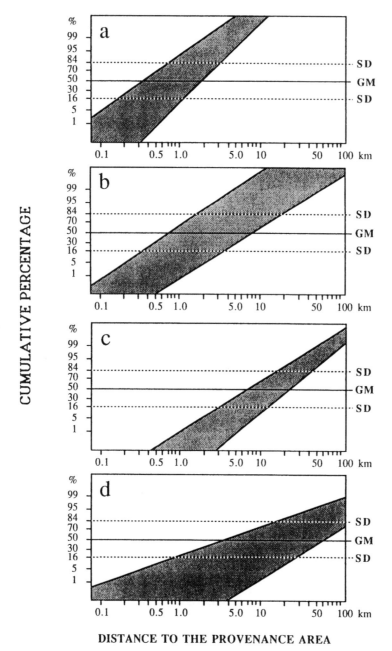

**DISTANCE TO THE PROVENANCE AREA**

Figure 7. The series of morainic landforms (A-D) and their generalized transport distance distributions. The data are based on 111 TDD-measurements in Finland (Salonen 1986). The transport distance of surface boulders increases from (fluted) hummocky moraines (A) to drumlin asssemblage (C) with a simultaneous decrease in their standard deviation. Mixed transport populations (D) in some ground moraine areas are characterized by variable transport distance and high value of standard deviation (gently sloping curves).

Table 2. Textural, structural, and dispersal characteristics of the various till facies of Central Québec.

| Till sheet | Facies | Fines (%) | Sorting | Structure | Transport characteristics K | %-Far-travelled | M |
|---|---|---|---|---|---|---|---|
| (diagram) | 4 | 5-20 | 1.3-2.2 | Numerous and varied | Not ob- served | 10-80% | 150 km |
| (diagram) | 3 | 15-30 | 2.5-3.0 | Lensed, bouldery | Few meters | < 5% | 5 km |
| (diagram) | 2 | 18-40 | 1.6-2.3 | Lense-stratified | 4 km | Variable, 5-80% | Not ob- served |
| (diagram) | 1 | 25-55 | 1.8-2.8 | Mostly massive, locally banded | 2 km | 5-35% | 15 km |

glacier finally stagnated during the last deglaciation. The underlying deposits from which the surface boulders are derived consists often of one or more lodgement till beds (Hirvas & Nenonen 1987); the boulders may then be derived from one or several 'till sheets', each one perhaps composed of a single layer of facies 1.

TRANSPORT DISTANCE AND GLACIAL FLOW PATTERN

Boulton and others (1985) described the glacier transporting system as a conveyor belt to which material is added by erosion and from which material is lost by deposition at many points along its length, but with a net discharge at the terminus, where material is dumped at the end of the conveyor belt. In addition to the depositional setting, Salonen (1986,1987) observed that the overall regularities in boulder transport statistics showed systematic variations from places to places within the Finnish part of the Fennoscandian Shield (Figure 8). These can be linked to both the glacier geometry and the glacial dynamic conditions during the deglaciation of Finland.

The length of the conveyor belt is a definite factor. For instance, near the location of the final ice divides, it appears that the glacier stagnated and melted away soon after the incorporation of the debris, and, consequently, a high content of local material is observed, as expected, whatever the morainic terrain (thick lines, Fig. 8). In contrast, in more peripheral areas relative to the final position of the divide, the boulders were transported over longer distances (Bouchard and Marcotte1986, Salonen 1986, Shilts et al. 1987). Although this appears trivial, it is nevertheless an important aspect of the boulder transport over both the Fennoscandian and the Canadian shield: since the transport distance seems to be a function of the size of the glacier, it is concluded that most of the boulders were eroded under the retreating ice sheet, while it was receding both in thickness and diameter.

The velocity of the glacier flow is a second important factor. By analogy with rivers, where an increase in velocity is associated with an increased sediment discharge (Drewry 1986), it is possible to suggest that within glacial lobes with very high basal sliding velocities substantial quantities of sediment will be transported further away by glaciers,

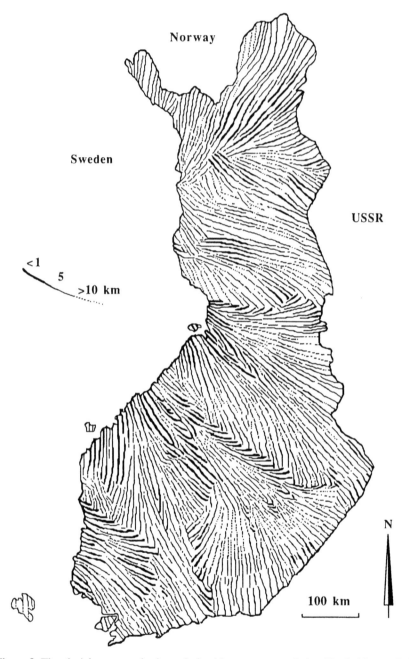

Figure 8. The glacial geometry is shown in boulder transport statistics. The boldness of glacial flow lines is inversely proportional to the value of geometric mean (GM) of transport distance of surface boulders (Salonen 1986, 1988).

as compared to areas where such lobate flow would not be formed (Clark 1987). An extreme case would be that of a surging glacier which would transport boulders over far greater distances than predicted by the facies and landform model discussed earlier. The average transport distance (GM) as mapped in Figure 8 is shown to be conformable to the pattern of rapid, lobate flow of the receding Scandinavian Ice Sheet as outlined by Punkari (1980). Thicker lines (shortest transport distances) are concentrated at the head and margins of the lobes.

### TRANSPORT OF BOULDERS AND DEGLACIATION OF FINLAND

Based on the regional variations in the characteristics of boulder transport in Finland, as discussed above, it can be safely assumed that many of the the compositional attributes of the surface drift were developed during the waning stages of the Scandinavian Ice Sheet. A major feature of the pattern of flow during these late-deglacial stages appears to be the development of numerous lobes with well defined transportation characteristics.

Based on the reconstruction of Punkari (1980), the surficial geological map of Kujansuu and Niemelä (1984) and the regional variations of the transport characteristics (Salonen 1986), a reconstruction of the waning Scandinavian Ice Sheet during the Late Weichselian in Finland is suggested as Figure 9.

According to Boulton and Jones (1979), there has been at least one surge operating in the basin of the Baltic Sea (Fig. 9) during the late Weichselian deglaciation. The ensuing rapid drainage of the continental glaciers destroyed their mass balance affecting the ice sheet well back from its marginal area, into the Shield areas. By analogy with the Laurentide Ice Sheet, it is suggested that these changes led to numerous new ice domes (Denton & Hughes 1981). Due to that, new ice flow regimes and patterns resulting from shifting centers of outflow, were formed. The pattern of the lobes was probably dependent of the thickness of the glacier (Minell 1980) at any one place, and on the rise of the proglacial lake/sea water level (Fig. 9), controlling the development and the rate of migration of calving bays.

### BOULDERS AND MINERAL EXPLORATION

The most spectacular discoveries of boulder tracing have probably already been made, at least in Finland, where about twenty mineable ores have been found during the last fifty years from that kind of prospecting. The times of mapping conspicuous boulder fans seem to be over, but there is still a strong demand for boulder tracing methods in modern exploration. Today, the search for ores is often based on weak, or isolated indicators showing possible subcropping of gold mineralization, or other targets seldomly producing traceable boulder trains or unambiguous geochemical anomalies.

In these cases, detailed till investigations are needed. The genesis of glacial deposits, their lithofacies, and transport characteristics have to be understood. If an ore indicator can be fitted into a particular landform/sediment association, its probable transport pathway can be reasonably well established using parameters such as K (distance to maximum abundance), D (half-distance), and GM (geometric mean of transport population) (Fig. 2).

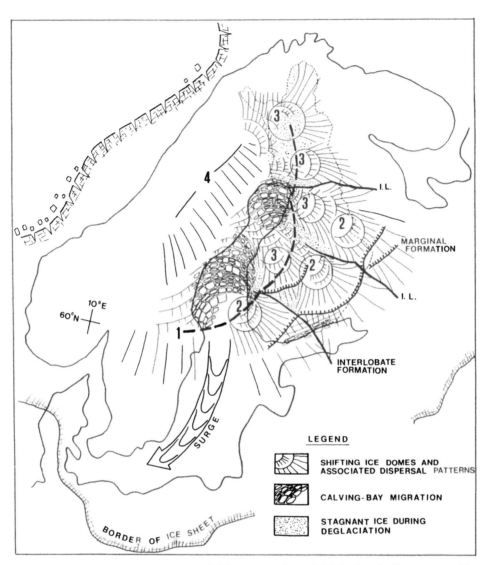

Figure 9. The disappearing of the Late Weichselian Ice Sheet in Finland and adjacent areas. The interpretation is based on boulder dispersal patterns (Fig. 3, Salonen 1987), and general map of Quaternary deposits of Finland (Kujansuu & Niemelä 1984). Successive positions 1 – 4 indicate the shifting of ice flow centers during deglaciation. The areas of more rapidly flowing ice lobes and the areas of passive ice, which prevailed during the deglaciation were determined by Punkari (1980). The calving ice front indicates the area with major glaciolacustrine sedimentation (area of waterlain tills). The Scandinavian Ice Sheet has been observed to have divided into several ice lobes during deglaciation. The lobes have their own flow regimes, and long belts of ice divides formed only in Sweden during the last phases of deglaciation (Lundqvist 1986).

CONCLUSION

The study of glacial boulders finds direct applications in mineral exploration and in the reconstruction of first-order flow features of ancient ice sheets with corollary informations on the changing ice sheet geometry and the successive positions of ice flow centers. The abundance of boulders in an area is primarily a geological factor. It reflects the ability of a given terraine to yield boulders from glacial erosion and the varying resistance of different lithologies to resist comminution during glacial transport.

All geological factors being equal, the transport distance of boulders varies systematically within a given area depending on the depositional setting, and from area to area, according to glacier flow conditions, and style of deglaciation. The till lithofacies model devised from the study of the anatomy of a single till sheet in Québec provides a useful concept explaining the variability of transport distances of boulders within a given area. The concept is applicable to the transport data of Finland. Till facies 1 contains upward an increasing proportion of distantly derived fragments, maximum transport distance (M) is around 15 km. It may form unfluted ground moraine or the upper surface of this lithofacies has been moulded into drumlinoid features. The average transport distance (GM) varies between 1–5 (ground moraine) and 7–20 km (drumlin areas). Lithofacies 2 is stratified, matrix supported diamicton of variably local to distal provenance. Lithofacies 3 is a massive, coarse grained diamicton of almost exclusive local derivation. It forms bouldery surfaces of active ice hummocky moraine with low value of average transport distance (GM = 0.5–2.0 km).

At regional scale the most important factors controlling the transport distances of boulders appear to be the size of the ice sheet and the development of lobate flow. Because most of the boulders are eroded under retreating ice sheet, while it was receding in diameter, the transport distance of boulders becomes shorter towards to the ice divide zones. In the core areas of continental glaciers, where glacial erosion has repeatedly been weak, the mixing of particles transported by several glacial cycles makes the boulder tracing studies more inconvenient.

REFERENCES

Bouchard, M. A. 1980. Late Quaternary geology, Témiscamie area, Central Québec, Canada. Ph.D. Dissertation, McGill University. 284 pp.
Bouchard, M.A. 1986. Géologie des dépôts meubles de la région de Témiscamie (Territoire-du-Nouveau-Québec). MM 83-03. Gouvernement du Québec. 80 pp.
Bouchard, M.A., Cadieux, B. & Goutier, Francoise 1984. L'origine et les charasteristiques des lithofacies du till dans le secteur nord du Lac Albanel, Québec: une étude de la dispersion glaciaire clastique. In: J. Guha & E.H. Chown (eds), Chibougamou-Stratigraphy and Mineralization. CIM Special Volume 34: 244-260.
Bouchard, M.A. & Martineau, G. 1984. Les aspects régionaux de la dispersion glaciaire, Chibougamou, Québec. In: J. Guha & E.H. Chown (eds), Chibougamou-Stratigraphy and Mineralization. CIM Special Volume 34: 431-440.
Bouchard, M.A. & Marcotte, C. 1986. Regional glacial dispersal patterns in Ungava, Nouveau-Québec. Current Research, Part B, Geological Survey of Canada, Paper 86-1B: 295-304.
Boulton, G.S. 1972. The role of thermal regime in glacial sedimentation. Institute of British Geographers Special Publication 4: 1-19.

Boulton, G.S. & Jones, A.S. 1979. Stability of temperate ice caps and ice sheets resting on beds of deformable sediment. Journal of Glaciology 24: 29-42.

Boulton, G.S., Smith, G.D, John, A.S. & Newsome, J. 1985. Glacial geology and glaciology of the last mid-latitude ice sheets. Journal of Geological Society, London 142: 447-474.

Clark, P.U. 1987. Subglacial sediment dispersal and till composition. Journal of Geology 95: 527-541.

Denton, G.H. & Hughes, T.J. 1981. The Arctic Ice Sheet: An Outrageneous Hypothesis. In: G.H. Denton & T.J. Hughes (eds), The Last Great Ice Sheets. John Wiley & Sons, New York: 437-467.

DiLabio, R. 1981. Glacial dispersal of rocks and minerals at the south end of Lac Mistassini, Quebec, with special reference to the Icon dispersal train. Geological Survey of Canada, Bulletin 323, 46 pp.

Drake, L.D. 1983. Ore plumes in till. Journal of Geology 91: 707-713.

Dreimanis, A. 1956. Tracing of Ore Boulders as a Prospecting Method in Canada. Canadian Mineralogical and Metallurgical Bulletin 51(550): 73-80.

Dreimanis, A. & Vagners, U.J. 1971. Bimodal distribution of rock and mineral fragments in basal till. In: R.P. Goldthwait (ed.), Till – A Symposium: 237-250. Ohio State University Press.

Drewry, D. 1986. Glacial Geologic Processes. Edward Arnold, London, 276 pp.

Dyke, A.S., Vincent, J.-S., Andrews, J.T., Dredge, L.A. & Cowan, W.R. 1987. The Laurentide Ice Sheet and an introduction to the Quaternary geology of the Canadian Shield. In: R.F. Fulton, J.A. Heginbottom & S. Funder (eds), Quaternary Geology of Canada and Greenland. Geological Survey of Canada, No. 1.

Elson, J.A. 1961. The geology of tills. In: E. Penner & J. Butler (eds.), Proceedings of 14th Canadian Soil Mechanics Conference, N.R.C., Association of Soil and Snow Mechanics, Technical Memoires 69.

Gillberg, G. 1965. Till distribution and ice movements of the northern slope of the South Swedish Highlands. Geologiska Föreningen i Stockholm Förhandlingar 86: 433-484.

Goldthwait, R.P. 1971. Introduction to Till, Today. In: R.P. Goldthwait (ed.), Till - A Symposium: 3-26. Ohio State University Press.

Grip. E. 1953. Tracing of glacial boulders as an aid to ore prospecting in Sweden. Economic Geology 48: 715-725.

Haldorsen, S. 1981. Grain-size distribution of subglacial till and its relation to glacial crushing and abrasion. Boreas 10: 91-105.

Hirvas, H. & Nenonen, K. 1987. Till stratigraphy of Finland. Geological Survey of Finland, Special Paper 3: 49-63.

Hughes, T.J. 1981. Numerical Reconstruction of Paleo-Ice Sheets. In: G.H. Denton & T.J. Hughes (eds), The Last Great Ice Sheets. John Wiley & Sons, New York: 221-261.

Hyvärinen, L., Kauranne, K. & Yletyinen, V. 1973. Modern boulder tracing in prospecting. Prospecting in areas of glaciated terrain 1973; Proceedings. IMM, London: 87-95.

Kinnunen, K.A. 1979. Ore mineral inclusions in detrital quartz contained in basal till and the glacial transport from the Ylöjärvi copper-tungsten deposit, southwestern Finland. Geological Survey of Finland, Bulletin 298, 55 pp.

Klassen, R.A. & Thompson, F.J. 1987. Glacial history and dispersal train configurations in Labrador. Geological Survey of Canada, Paper 87-8.

Krumbein, W.C. 1937. Sediments and exponential curves. Journal of Geology 45: 577-601.

Kujansuu, R. & Niemelä, J. 1984. The Quaternary deposits in Finland, map 1:1 000 000. Geological Survey of Finland.

Lawson, D.E. 1981. Sedimentological characteristics and classification of depositional processes and deposits in the glacial environment. CRREL Report 81-27, 16 pp.

Lee, H.A. 1965. Investigation of eskers for mineral exploration. Geological Survey of Canada, Paper 65014. 17 pp.

Lundqvist, J. 1969. Beskrivning till jordartskarta över Jämtlands län. Sveriges Geologiska Undersökning, Serie Ca, 45. 418 pp.

Lundqvist, J. 1986. Late Weichselian Glaciation and Deglaciation in Scandinavia. In: V. Sibrava, D.Q. Bowen & G.M. Richmond (eds), Quaternary glaciations in the Northern Hemisphere: Quaternary Science Reviews 5: 269-292. Pergamon Press, Oxford.

Marcussen, Ib 1973. Stones in Danish tills as a stratigraphical tool. A review. Bulletin of the Geological Institute of the University of Uppsala, New Serie 5: 177-181.

Minell, H. 1980. The distribution of local bedrock material in some moraine forms from the inner part of northern Sweden. Boreas 9: 275-281.

Peltoniemi, H. 1985. Till lithology and glacial transport in Kuhmo, eastern Finland. Boreas 14: 67-74.

Perttunen, M. 1977. The lithologic relation between till and bedrock in the region of Hämeenlinna, southern Finland. Geological Survey of Finland, Bulletin 291, 68 pp.

Prest, V.K. & Nielsen, E. 1987. The Laurentide Ice Sheet and long-distance transport. Geological Survey of Finland, Special Paper 3: 91-101.

Punkari, M. 1980. The ice lobes of the Scandinavian ice sheet during the deglaciation in Finland. Boreas 9: 307-310.

Richard, P., LaRouche, A. & Bouchard, M.A. 1981. Age de la déglaciation finale et histoire postglaciaire de la végétation dans la partie centrale du Nouveau-Québec. Géographie Physique et Quaternaire XXXVI: 63-90.

Salminen, R. 1980. On the geochemistry of copper in the Quaternary deposits in the Kiihtelysvaara area, North Karelia, Finland. Geological Survey of Finland, Bulletin 309, 48 pp.

Salonen, V.-P. 1986. Glacial transport distance distribution of surface boulders in Finland. Geological Survey of Finland, Bulletin 338, 57 pp.

Salonen, V.-P. 1987. Observations on boulder transport in Finland. Geological Survey of Finland, Special Paper 3: 103-110.

Salonen, V.-P. 1988. Application of glacial dynamics, genetic differentiation of glacigenic deposits and their landforms to indicator tracing in the search for ore deposits. In: R.P. Goldthwait & C. Matsch (eds), Genetic Classifications of Glacigenic Deposits and Their Landforms. Balkema, Rotterdam: 183-196.

Saltikoff, B. 1984. Boulder tracing and the mineral indication data bank in Finland. Prospecting in areas of glaciated terrain 1984; Proceedings: IMM, London: 179-191.

Sauramo, M. 1924. Tracing of glacial boulders and its application in prospecting. Bulletin de la Commission géologique de Finlande 67, 37 pp.

Shilts, W.W. 1973. Glacial Dispersal of Rocks, Minerals and Trace Elements in Wisconsinan Till, Southeastern Quebec, Canada. Geological Society of America, Memoir 136: 189-219.

Shilts, W.W. 1976. Glacial till in mineral exploration. In: R.F. Legget (ed.), Glacial Till. An Interdisciplinary Study. The Royal Society of Canada, Special publication 12: 205-224.

Shilts, W.W. 1985. Geological models for the configuration, history and style of disintegration of the Laurentide Ice Sheet. In: M.J. Woldenberg (ed.), Models in Geomorphology. State University of New York, Buffalo. Allen & Unwin, Boston: 73-91.

Shilts, W. W., Cunningham, C.M. & Kaszycki, C.A. 1979. Keewatin Ice Sheet – Re-evaluation of the traditional concept of the Laurentide Ice Sheet. Geology 7: 537-541.

Shilts, W.W., Aylsworth, J.M., Kaszycki, C.A., & Klassen, R.A. 1987. Canadian Shield. In: W.L. Graf (ed.), Geomorphic systems of North America,. Geological Society of America, Centennial Special Volume 2: 119-161.

Wentworth, C.K. 1922. A scale of grade and class terms for clastic sediments. Journal of Geology 30: 377-392.

CHAPTER 7

# Glacial dispersal trains*

R. N. W. DILABIO
*Geological Survey of Canada, Ottawa, Ontario*

INTRODUCTION

Exploration of glaciated terrain by means of indicator tracing would be simplified greatly if the exploration geologist or prospector had some general guidelines on what patterns of glacial dispersal he could expect to find in a given area. Several papers in this volume explain methods for determining the ice-flow direction and the distance of transport of glacially transported debris. This paper will describe the main features of dispersal trains and illustrate several in order to show what features they have in common, as well as the variability in those features.

CHARACTERISTICS OF DISPERSAL TRAINS

What characteristics do we see in examining dispersal trains that have been traced back to their sources? Miller (1984) has published a model (Fig. 1) that is applicable. Based on observations of small trains that were studied during mineral exploration, this model shows a plan view and cross-sections through a hypothetical train. Another model that applies is that of Drake (1983). Small dispersal trains have at least five features in common. First, they are extremely thin in comparison to their length and width. Second, they are really hundreds to thousands of times larger than their bedrock sources, hence they form large targets for geochemical and lithological exploration. Third, they have abrupt lateral and vertical contacts with the enclosing till. Fourth, they climb gently within the enclosing till as they are followed down-ice from their sources. Fifth, many, but not all, show a down-ice decrease in content of distinctive components from the 'head' to the 'tail' of the train (Fig. 2; Shilts, 1976). For some years in Finland (Salonen, 1986) and recently in North America (Clark, 1987; Strobel and Faure, 1987), Krumbein's (1937) concept of half-distance (analogous to the half-life of a radioisotope) has been applied to these decay curves in order to estimate transport distance from isolated data and to interpret the mechanisms of debris transport.

The tail of a dispersal train is generally many times longer than the head and is usually the part of the train that is detected first by mineral exploration programs using till geochemistry or boulder tracing. Dispersal trains of distinctive boulders, minerals, trace

*Geological Survey of Canada, Contribution No. 13288.

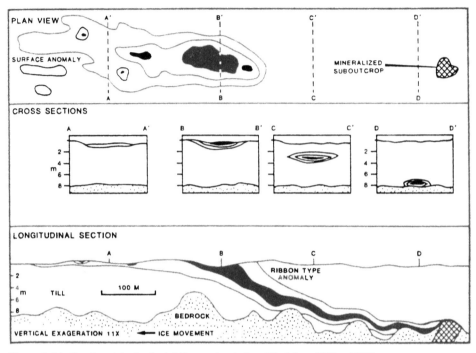

Figure 1. Idealized model of a glacial dispersal train (modified from Miller, 1984).

elements or major elements, and radioactive components may increase the size of mineral exploration targets by several orders of magnitude. The major objective of indicator tracing is simply to detect the tail of a dispersal train, trace it back to its head, and find its source.

At a given site, the composition of till may be the composite of many overlapping dispersal trains. This blending of debris derived from different up-ice sources either along one flowline or from shifting flowlines produces the mixed lithology that is a normal property of till. Most of the individual overlapping 'trains' cannot be identified, however, because they are too small or are composed of rocks or minerals that are not distinctive. The size and shape of a dispersal train are controlled by the orientation of the source relative to ice flow, by the size and erodibility of source, and by the influence of topography on ice flow in the source and dispersal areas, which can trap trains in valleys or break them into disjointed segments in rough terrain.

Dispersal occurs at a variety of scales ranging from continental (100's of kilometres), to regional (100 to 10's of kilometres), to local (<10 kilometres), to property-scale (final stages of mineral exploration in the 100's to 10's of metres) (Shilts, 1984). Dispersal trains exist that are up to hundreds of kilometres in length, such as the train of carbonate-rich till that extends southwards from Hudson Bay in Canada (Shilts et al., 1979). Trains of this size are detected only where the characteristic lithological component of the train is present in adequate amounts and is distinctive against the background rock types in the dispersal area. For drift prospecting purposes, these large trains are

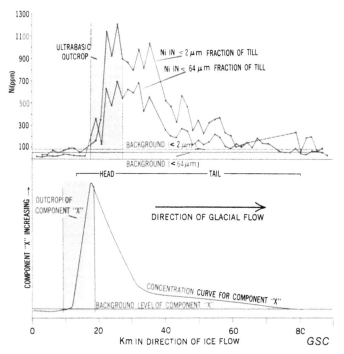

Figure 2. Dispersal curves for a distinctive component in till. Actual data for nickel in till are from the Thetford Mines area, Quebec. Idealized curves (bottom) show the relationship of the head and tail of a negative exponential curve (after Shilts, 1976).

significant in that the exotic lithology of the till can mask the lithology and geochemistry of mineralized debris eroded from local sources (Geddes and Kristjansson, 1986; Kaszycki and DiLabio, 1986; Gleeson and Sheehan, 1987; Karrow and Geddes, 1987). These trains can be detected by 'reconnaissance' scale sampling of till – on the order of one sample per 100 km². Other media, such as lake sediments, can also be used to detect them (Coker and Shilts, 1979).

Smaller dispersal trains derived from individual rock units, distinctive belts of rock, or scattered mineralization are more likely to be detected in the preliminary stages of mineral exploration programs. At this stage of exploration, 'local' scale sampling on the order of one till sample per km², will be enough to define which parts of a favourable bedrock unit are most metalliferous and may even detect the tails of dispersal trains derived from small sources. At this sampling density, any large trains that are identified will focus interest on areas that should be sampled at a detailed scale to clarify whether the large trains represent large areas of mineralized bedrock or simply represent areas of high background metal levels or are composed of several overlapping small trains derived from scattered areas of mineralized bedrock.

'Detailed' sampling, in which sample spacing is 10 to 100 metres, is designed to locate the heads of dispersal trains, and ultimately, their sources. This density of sampling would normally be carried out in tracing trains up-ice to their sources or in testing geophysical anomalies and geological structures or contacts that are known to be

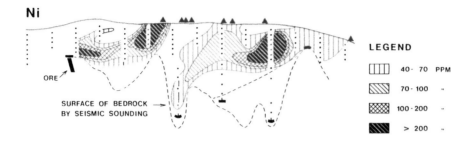

**Ni**

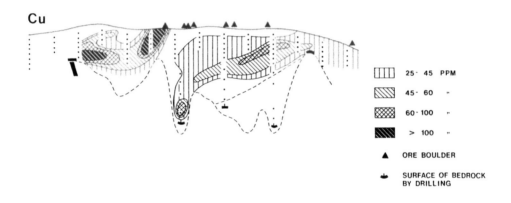

**Cu**

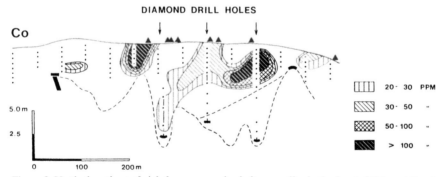

**Co**

Figure 3. Vertical sections of nickel, copper, and cobalt anomalies in the fine (<0.05 mm) fraction of till at Tervo, Finland (after Nurmi, 1976).

favourable for mineralization. At this scale of sampling, postglacial mobilization of boulders by solifluction and of trace elements in till by groundwater flow may become apparent. These postglacial events spread the dispersal train downslope, partially obscuring its original shape, which was the result of clastic dispersal (e.g., DiLabio et al., 1982).

EXAMPLES OF DISPERSAL TRAINS

Many dispersal trains have been described in the past few years, particularly in Fenno-scandia and Canada (Minell, 1978; Bolviken and Gleeson, 1979; Hyvärinen et al., 1973; Salonen, 1986; Puranen, 1988; Coker and DiLabio, 1989). The dispersal trains chosen for this paper illustrate most of their characteristics and are mainly small ones that have been mapped by 'detailed' sampling of small exploration properties.

It has been clear for a long time that only through careful studies of the internal structure ('anatomy') of dispersal trains can the exploration geologist learn how to use trains to find mineralized bedrock. As long ago as 1924, Sauramo was describing the dispersal patterns of ore boulders in Finland. More recently, Kauranne (1959) and Wennervirta (1968) continued those studies by publishing landmark papers that signalled the start of a major expansion of research work in the 1970's and 1980's.

One of the most detailed and valuable of the modern studies was carried out by Nurmi (1976), who performed a three-dimensional dissection of the dispersal train derived from the Talluskanava nickel-copper deposit at Tervo, Finland. Sampling of the train was first done every 20 m on lines 100 m apart; by the end of the project, the level of detail had increased to the point where the faces of test pits had been sampled on a 20 cm by 20 cm grid. In one pit, sequential slices of the till face were made, resulting in a pattern of samples taken every 20 cm in all three directions in a cube of till 2 m on a side. This work showed conclusively that the dispersal train had very sharp contacts with the enclosing barren till; trace element levels dropped across a distance of 40 cm or even <20 cm from highly anomalous to background levels as the edge of the train was crossed in section. If the effect of hydromorphic redistribution of the metals is considered, then the contacts must be extremely sharp.

Similar work on sections parallel to ice-flow produced longitudinal sections (Fig. 3) that show the shape of the dispersal train as it extends down-ice for a distance of at least 550 m. Elevated levels of nickel, copper, and cobalt all portray the dispersal train as consisting of two distinct lenses of metal-rich till that are 2 to 3 m thick and that dip up-ice at a very low angle.

In another project in Finland, Salminen and Hartikainen (1985) determined the patterns of glacial dispersal from mineralized bedrock and distinctive rock units at 11 localities in North Karelia. Samples were collected in test pits and with light drills at sites along the long axes of dispersal trains. Analysis of their pebble lithology and trace element geochemistry showed again that at many of the localities, the dispersal train was a thin, lithologically distinct lens of till that dipped at a very low angle up-ice towards its bedrock source.

In the Strange Lake area of Labrador, a large dispersal train of debris derived from a peralkaline granite complex containing REE, Nb, Zr, Y, Be, and Li has been described by McConnell and Batterson (1987). This dispersal train was detected originally by a U anomaly in a regional reconnaissance lake sediment survey (Geological Survey of Canada, 1979). Boulder tracing and an airborne radiometric survey eventually led to the discovery of the mineralized granite. It is important to note that depending on the distinctiveness of a source, for example its radioactivity, and the amount of bedrock exposure and/or overburden thickness, remote techniques or other regional recon-naissance methods can detect the larger dispersal trains. The Strange Lake train is over 30 km long, 1 to 2 km wide, and no more than 5 m thick. It maintains its width for about 20

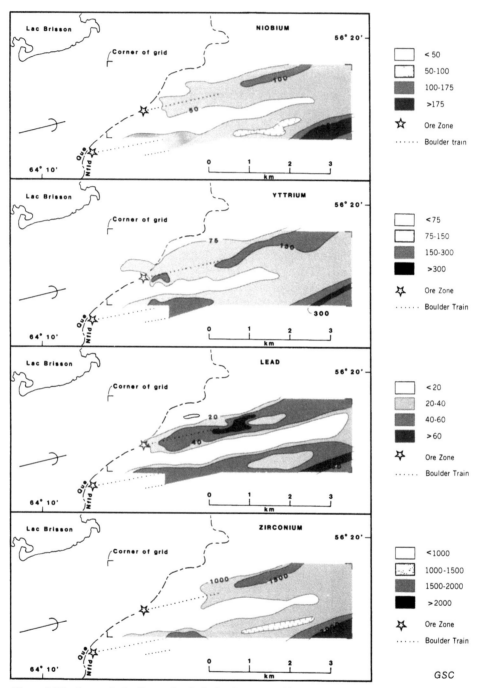

Figure 4. The Strange Lake dispersal train, Labrador, mapped by the abundance of trace elements in the <63 μm fraction of till.

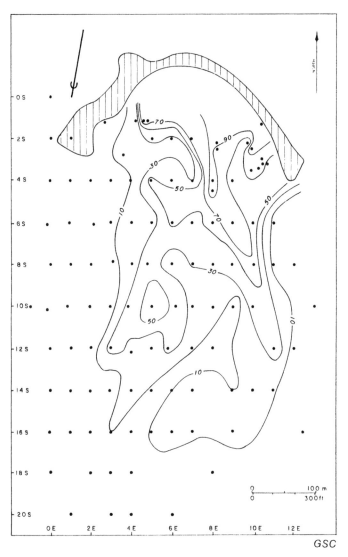

Figure 5. Contoured map of the abundance of ore pebbles (4 to 64 mm) in till in the Icon dispersal train, Quebec.

km. Two coalescent trains that comprise its head (Fig. 4) are identifiable in the boulder fraction of the till on the surface of the tail of the train at great distances from the source because they were derived from slightly different phases of the granite. In its tail, the train goes over bedrock obstacles with 100 m of relief but does not change orientation.

Detailed sampling (sampling interval of 100 m on lines 400 m apart) in the head of the train found that Pb, Nb, Y, and Zr contents of the till best defined and outlined the details within the head. At the source, only the lower metre of the till, where it rests on the down-ice edge of the ore subcrop, is ore-bearing. The train reaches the surface a short

Figure 6. Close-up of the base of the Icon train showing the sharp contact between the copper-rich till and the underlying sediments. Ice movement was away from the viewer. GSC photo 203875-Q.

Figure 7. Longitudinal section in the Icon train showing the dark-coloured copper-rich till rising over pale-coloured sands. Ice movement was from right to left. GSC photo 203818.

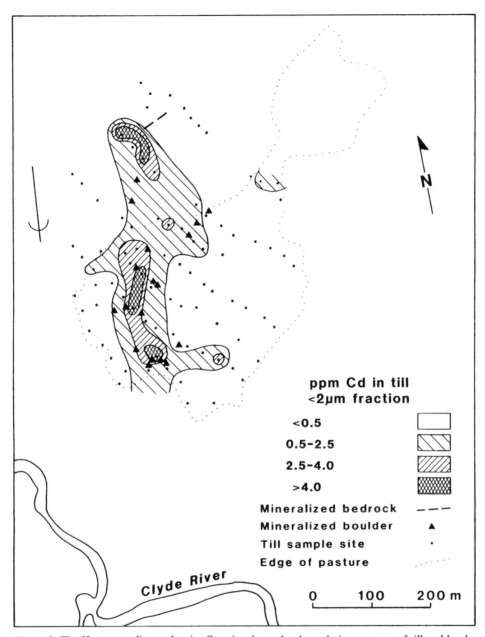

Figure 8. The Hopetown dispersal train, Ontario, shown by the cadmium content of till and by the distribution of ore boulders.

distance (about 100 m) down-ice. The sides of the train are abrupt, marked by a change from background to highly anomalous levels of Nb, Y, Pb, and Zr over a few tens of metres where a lateral edge is crossed.

The Icon train (Fig. 5) in central Quebec, derived from a high grade vein of copper ore (DiLabio, 1981), is about 650 m long and 300 m wide. In sections through the train, the ore-bearing till truncates the structure of older ice-contact sediments and ablation till (Fig. 6). Very little mixing with older sediments took place, for the till is almost as rich in copper as is the *in situ* ore; it was actually mined as ore. In the tail of the train, the copper-rich till is in places only 10 centimetres thick, at the top of the soil profile, but was recognizable because of its unusual composition. In longitudinal section, the train rises over its sandy substrate at an angle of about 10° through the first 100 m of its length, then at a lower angle (Fig. 7). How much of the underlying sediment was eroded is unknown, but down-ice dilution of the copper-rich till is minimal. In three dimensions, the train is shaped like the inverted bowl of a spoon, pointing down-ice. The attitude of this train is controlled by the shape of the underlying sediments, and it is the surface till, so it is not a good example of a train rising down-ice within a till sequence. The Icon train is, however, one of the better examples of the sharpness of the lower contact of a train and of the lack of mixing with underlying sediments.

The Hopetown train (DiLabio et al., 1982), in southern Ontario, was derived from a small zinc occurrence in marble. Zinc distribution in the till is influenced too much by postglacial remobilization to be useful in mapping the dispersal train, but cadmium (Fig. 8) released during weathering of the sphalerite is much less mobile and reflects only clastic transport. The distribution of ore boulders matches the cadmium pattern. Aligned parallel to local striae, the train is straight, has sharp edges, and is more than 400 m long and consistently about 100 m wide. The till is less than 10 m thick and the train is continuous over its full extent (from trenching).

An overburden drilling program led to the discovery of the EP gold zone at Waddy Lake, Saskatchewan (Averill and Zimmerman, 1986). The authors mapped a dispersal train (Fig. 9) in which heavy mineral concentrates from the till contained abundant gold, electrum, copper, galena, chalcocite, and pyromorphite. This classic dispersal train is ribbon-shaped, with sharp edges, and is at least 600 m long and 50 to 100 m wide. Of note is the unexplained 15° difference between the orientation of the train and the orientation of local striae. It is possible that the striae reflect a late glacial ice flow direction that differed from the earlier direction that produced the train, a widespread phenomenon on the Canadian Shield (Veillette, 1986; Kaszycki and DiLabio, 1986; Klassen and Thompson, 1987).

Near Timmins, Ontario, an overburden drilling program was carried out by Bird and Coker (1987) at the Owl Creek gold mine. The drilling revealed a thick and complex Quaternary record with up to four glacial sediment units, each having a different ice movement direction that was assigned to each till after till fabric and striae studies were done in the open pit and around the mine (Fig. 10). In the lowest ('Older') till, which rests on the bedrock, dispersal is localized because the till pinches out against a bedrock ridge. The highest gold levels in this till are adjacent to the subcropping gold mineralization. The overlying (Matheson) till has not been in direct contact with mineralized bedrock, but has derived its gold by recycling the much older (possibly pre-Wisconsinan, DiLabio et al ., 1988) lower till. Dispersal in the upper till is longer, about 600 m, and its area of maximum gold content, which could be called its head, is displaced about 300 m

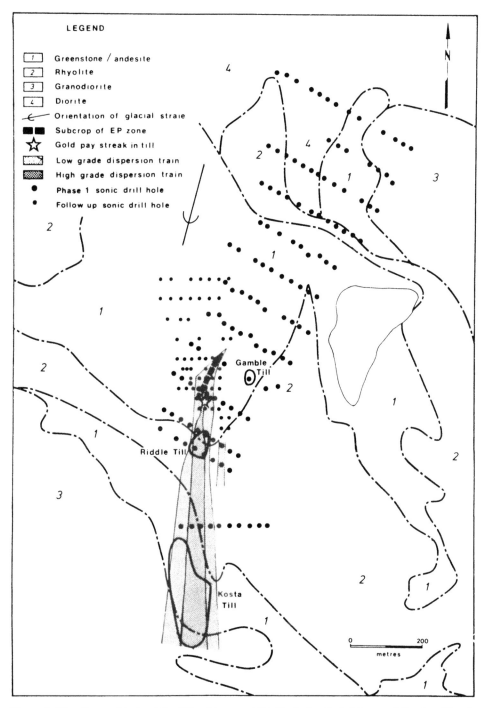

Figure 9. The dispersal train at the EP gold zone at Waddy Lake, Saskatchewan (after Averill and Zimmerman, 1986).

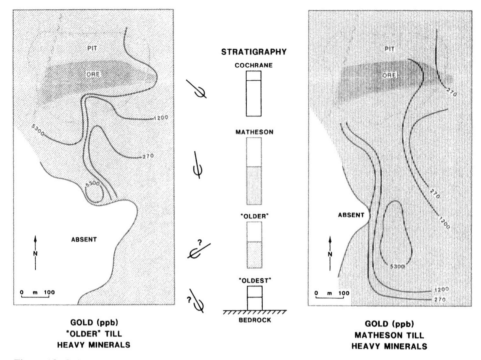

Figure 10. Gold abundances in heavy mineral concentrates from the 'Older' and Matheson tills at the Owl Creek gold mine, Timmins, Ontario (after Bird and Coker, 1987).

down-ice from the subcropping gold mineralization. This example emphasizes the effect of bedrock topography and recycling on dispersal and shows the importance of understanding the glacial stratigraphy and glacial flow history in any region under study.

CONCLUSIONS

Detailed studies of dispersal trains allow their three-dimensional shapes and structures to be determined. Many trains are very thin in comparison to their length and width (T:W:L ratios on the order of 1:200:1000 are common). They also characteristically have straight edges, constant widths, abrupt contacts with enclosing till, and gentle angles of climb as they are followed down-ice from their bedrock sources. The following characteristics can be helpful to the geologist or prospector. First, dispersal trains are much larger than their bedrock sources, making them easier targets to find. Second, they are usually straight, oriented parallel to one ice-flow direction, so they can be followed up-ice to the source. Third, they climb gently (1° to 3°) in the down-ice direction, so they are not prone to sudden changes in attitude or depth, except where they pinch out against substrate obstructions or where they have been eroded away. Fourth, they are often very thin, which forces the geologist to sample short intervals in sections and drill core so that a train is not diluted artificially by combining barren till with ore-bearing till during sampling.

REFERENCES

Averill, S.A. and J.R. Zimmerman 1986. The Riddle Resolved: The Discovery of the Partridge Gold Zone using sonic drilling in glacial overburden at Waddy Lake, Saskatchewan; Canadian Geology Journal of the Canadian Institute of Mining and Metallurgy 1: 14-20.

Bird, D.J. and W.B. Coker 1987. Quaternary stratigraphy and geochemistry at the Owl Creek gold mine, Timmins, Ontario, Canada. In: R.G. Garrett, (ed.), Geochemical Exploration 1985, Part 1. Journal of Geochemical Exploration 28: 267-284.

Bolviken, B. and C.F. Gleeson 1979. Focus on the use of soils for geochemical exploration in glaciated terrane. In: P.J. Hood (ed), Geophysics and Geochemistry in the search for Metallic Ores. Geological Survey of Canada, Economic Geology Report 31: 295-326.

Clark, P.U. 1987. Subglacial sediment dispersal and till composition. Journal of Geology 95: 527-541.

Coker, W.B. and R.N.W. DiLabio 1989. Geochemical exploration in glaciated terrain: geochemical responses. Exploration '87, Ontario Geological Survey, Special Volume 3: 336–383.

Coker, W.B. and W.W. Shilts 1979. Lacustrine geochemistry around the north shore of Lake Superior: implications for evaluation of the effects of acid precipitation. Current Research, Part C, Geological Survey of Canada, Paper 79-1C: 1-15.

DiLabio, R.N.W. 1981. Glacial dispersal of rocks and minerals at the south end of Lac Mistassini, Quebec, with special reference to the Icon dispersal train. Geological Survey of Canada, Bulletin 323, 46 pp.

DiLabio, R.N.W., R.F. Miller, R.J. Mott and W.B. Coker 1988. The Quaternary stratigraphy of the Timmins area, Ontario, as an aid to mineral exploration by drift prospecting. Current Research, Part C, Geological Survey of Canada, Paper 88-1C: 61-65.

DiLabio, R.N.W., A.N. Rencz and P.A. Egginton 1982. Biogeochemical expression of a classic dispersal train of metalliferous till near Hopetown, Ontario. Canadian Journal of Earth Sciences 19: 2297-2305.

Drake, L.D. 1983. Ore plumes in till. Journal of Geology 91: 707-713.

Geddes, R.S. and F.J. Kristjansson 1986. Quaternary geology of the Hemlo area: constraints on mineral exploration. Canadian Geology Journal of the Canadian Institute of Mining and Metallurgy 1(1): 5-8.

Geological Survey of Canada 1979. Regional lake sediment and water geochemical reconnaissance data, Labrador. Geological Survey of Canada, Open File 559.

Gleeson, C.F. and D.G. Sheehan 1987. Humus and till geochemistry over the Doyon, Bousquet, and Wiliams gold deposits. Canadian Institute of Mining and Metallurgy Bulletin 80 (898): 58-66.

Hyvärinen, L., K. Kauranne and V. Yletyinen 1973. Modern boulder tracing in prospecting. In: M.J. Jones (ed.), Prospecting in Areas of Glaciated Terrain – 1973 Institution of Mining and Metallurgy, London, 87-95.

Karrow, P.F. and R.S. Geddes 1987. Drift carbonate on the Canadian Shield. Canadian Journal of Earth Sciences 24: 365-369.

Kaszycki, C.A. and R.N.W. DiLabio 1986. Surficial geology and till geochemistry, Lynn Lake-Leaf Rapids region, Manitoba; Current Research, Part B, Geological Survey of Canada, Paper 86-1B: 245-256.

Kauranne, L.K. 1959. Pedogeochemical prospecting in glaciated terrain. Commission geologique de Finlande, Bulletin 184: 1-10.

Klassen, R.A. and F.J. Thompson 1987. Ice flow history and glacial dispersal in the Labrador Trough. Current Research, Part A, Geological Survey of Canada, Paper 87-1A: 61-71.

Krumbein, W.C. 1937. Sediments and exponential curves. Journal of Geology 45: 577-601.

McConnell, J.W. and M.J. Batterson 1987. The Strange Lake Zr-Y-Nb-Be-REE deposit, Labrador: a geochemical profile in till, lake and stream sediment, and water. In: R.G. Garrett (ed.), Geochemical Exploration 1985. Journal of Geochemical Exploration 29: 105-127.

Miller, J.K. 1984. Model for clastic indicator trains in till. Prospecting in Areas of Glaciated Terrain – 1984. Institution of Mining and Metallurgy, London: 69-77.

Minell, H. 1978. Glaciological interpretations of boulder trains for the purpose of prospecting in till. Sveriges Geologiska Undersökning, Series C, no. 743, 51 pp.

Nurmi, A. 1976. Geochemistry of the till blanket at the Talluskanava Ni-Cu ore deposit, Tervo, Central Finland. Geological Survey of Finland, Report of Investigation No. 15, 84 pp.

Puranen, R. 1988. Modelling of glacial transport of basal tills in Finland. Geological Survey of Finland, Report of Investigation 81, 36 pp.

Salminen, R. and A. Hartikainen 1985. Glacial transport of till and its influence on interpretation of geochemical results in North Karelia, Finland. Geological Survey of Finland, Bulletin 335, 48 pp.

Salonen, V-P. 1986. Glacial transport distance distributions of surface boulders in Finland. Geological Survey of Finland, Bulletin 338, 57 pp.

Sauramo, M. 1924. Tracing of glacial boulders and its application in prospecting. Commission geologique de Finlande, Bulletin 67, 37 pp.

Shilts, W.W. 1976. Glacial till and mineral exploration. In: R.F. Legget (ed.), Glacial Till: an Interdisciplinary Study. Royal Society of Canada, Special Publication 12: 205-224.

Shilts, W.W. 1984. Till geochemistry in Finland and Canada. Journal of Geochemical Exploration 21: 95-117.

Shilts, W.W., C.M. Cunningham and C.A. Kaszycki 1979. Keewatin ice sheet-reevaluation of the traditional concept of the Laurentide ice sheet. Geology 7: 537-541.

Strobel, M.L. and G. Faure 1987. Transport of indicator clasts by ice sheets and transport half-distance: a contribution to prospecting for ore deposits. Journal of Geology 95: 687-697.

Veillette, J.J. 1986. Former southwesterly ice flows in the Abitibi-Timiskaming region: implications for the configuration of the late Wisconsinan ice sheet. Canadian Journal of Earth Sciences 23: 1724-1741.

Wennervirta, H. 1968. Application of geochemical methods to regional prospecting in Finland. Commission geologique de Finlande, Bulletin 234, 91 pp.

# CHAPTER 8

# Compositional variability of till in marginal areas of continental glaciers

ROBERT A. STEWART
*HRP Associates, Inc., Plainville, Conn., USA*

BRUCE E. BROSTER
*Department of Geology, University of Windsor, Ontario, Canada*

## INTRODUCTION

When indicator erratics occur in sufficient concentrations in moraines and till sheets, their geographic distribution in a dispersal train or fan may be sufficient to locate their source, without having to know much about till as a transport medium. In cases where indicator erratics are sparse, or are restricted to a particular till unit which may or may not crop out, genetic analysis of the till may be necessary to ascertain the transport history and provenance of the indicators. Although the complex marginal glacial environment may not be very suitable for indicator tracing, the composition of till and related sediments found there can still yield valuable information about the general glacial provenance of a region.

The chapter focuses on compositional variations in tills and related sediments occurring in three settings commonly found in marginal areas of continental glaciers:

1. A head-of-outwash glaciodeltaic environment in Cape Cod, Massachusetts that consists of outwash and subaquatic glaciogenic diamictons.

2. Intra-till variations within a thick (25 m) exposure of the late Wisconsinan St. Joseph Till near Port Albert, southern Ontario.

3. Inter-till variations among basal till sheets in Iowa that range in age from Pre-Illinoian to late Wisconsinan.

## LANDFORM-SEDIMENT ASSOCIATIONS IN MARGINAL AREAS OF CONTINENTAL GLACIATION

### Moraines and related landforms

End moraines and recessional moraines consisting principally of till delimit, respectively, glacial advances, and readvances or stillstands (Fig. 1A-D). Mickelson et al. (1983) noted that morainal landforms were either true end moraines (simple, superposed or rock-cored) or palimpsests (Fig. 1). Simple end moraines (Fig. 1A) consist chiefly of till formed at an ice margin during the last episode of glaciation. The thickness of the youngest till in the ridge is usually significantly greater than the thickness in the upglacial direction. Simple end moraines form by thickening of till during an advance, and their topographic form reflects only the increased thickness of the till. Superposed end moraines (Fig. 1B) owe part of their form and relief to the configuration of an underlying,

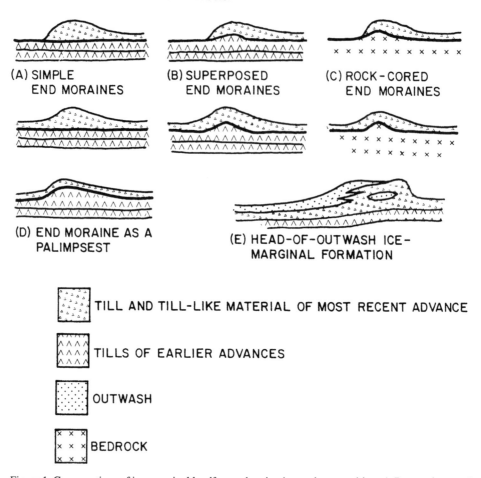

Figure 1. Cross sections of ice-marginal landforms showing internal composition: A-D, moraines; and E, head-of-outwash moraine-like ridge (after Mickelson et al., 1983).

overridden moraine. As with other varieties of true moraines, the till thickness in a superposed end moraine is greater than in an upglacial direction, and multiple moraine crests are common. Rock-cored end moraines (Fig. 1C) are similar to simple end moraines in terms of an increased till thickness; however, the till is deposited atop a bedrock ridge that stabilized the ice margin. Palimpsests (Fig. 1D) are not true end moraines, but consist of a till veneer draping an older end moraine, which may have been partly eroded during the later event. Unlike other examples of true moraines (Fig. 1A-C), palimpsests do not reflect stabilization of the glacier margin and thickened accumulations of till.

End moraines deposited under terrestrial conditions typically comprise two facies associations: (1) a subglacial sediment facies consisting of basal tills with occasional small, local lenses of sorted sediments, and (2) an overlying supraglacial sediment association that contains abundant glaciofluvial sediments and till-like diamictons (Boulton, 1972; Lawson, 1979; Kemmis et al., 1981). The local stratigraphic relationships between these two sediment facies may be complex because of minor fluctuations of the glacier margin (Lawson, 1979).

In many areas of North America, Europe, and Scandinavia, regional retreat of continental glaciers was accompanied by the development of large proglacial lakes, or the incursion of marine waters. Moraines constructed under these conditions are commonly little more than heads-of-outwash, and may consist of predominantly outwash with lesser amounts of till or till-like material (Fig. 1E; cf. Evenson and Clinch, 1987).

Because of the diverse sources and methods of delivery of debris at glacier termini, the composition of till and related sediments in heads-of-outwash and true end moraines is likely to consist of both local and distal debris. Therefore, in both subaerial and subaquatic environments, these landform associations should be viewed as the most variable and least useful zones for indicator tracing in a particular glaciated region. Ore boulders and geochemical anomalies in moraines should be treated as indications of other, similar, phenomena that may occur in till sheets behind the moraines.

### Ground moraine (till sheet)

Ground moraine consisting of basal till is the principal glaciogenic sediment sought and examined by explorationists, as it is generally considered to be the most useful medium for the delineation of source areas of glacial indicator clasts. Difficulties in tracing a particular component in till during exploration programs may often stem from a failure to appreciate the degree and cause of compositional variability in tills. Till variability may be either (1) a function of variability within a single till sheet (intra-till variability), or (2) a function of the differences between till sheets (inter-till variability).

Intra-till variability is considered to be due to differences in materials encountered along the transport path, glacial dynamics, and mode of transport and deposition (basal or supraglacial). It is commonly assumed that intra-till variability increases toward the glacial margin as more subglacial units are encountered and intra-ice dispersion is increased with increasing transport distance (Anderson, 1957). In most glaciers, however, basal shearing, regelation, thrusting and alternating deposition and re-incorporation of debris can result in homogenization of till composition for some distance behind an end moraine (Kemmis, 1981; Broster, 1986). Inter-till variability is mainly considered to be due to differences in materials incorporated along the glacial flow path, which may include bedrock, older till sheets, loess and paleosols (Knightly and Stewart, 1986).

### COMPOSITION OF GLACIAL DRIFT AT A HEAD-OF-OUTWASH ICE SHEET MARGIN, CAPE COD, MASSACHUSETTS

The physiographic region of eastern Massachusetts that includes Cape Cod, Martha's Vineyard and Nantucket is the product of three lobes of the Wisconsinan ice sheet which were active during deglaciation of the region (Fig. 2). From west to east, they are named the Buzzards Bay lobe (Schafer, 1980), the Cape Cod Bay lobe and the South Channel lobe (Woodworth and Wigglesworth, 1934). Retreat of these lobes was not synchronous, as various stratigraphic evidences show the deposits of the Buzzards Bay lobe to be older than the deposits of the Cape Cod Bay lobe (Mather et al., 1942), which are in turn older

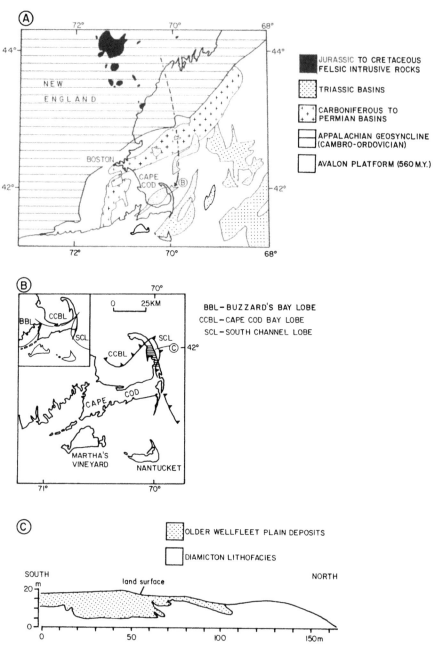

Figure 2. (A) Regional geology of New England and Cape Cod region (after Ballard and Uchupi, 1975). Dashed line shows generalized path of South Channel lobe toward Wellfleet area (B), based on glacial erratics, geomorphic and stratigraphic considerations. (B) Cape Cod region showing position of ice margin (sawtooth line) at time of deposition of glaciodeltaic sediments (C) in the Wellfleet Quadrangle (horizontal lines) (after Oldale, 1982). Inset in (B) shows generalized lobate configuration of glacial margin prior to deposition of Wellfleet outwash plain. (C) Profile of sea cliffs at Newcomb Hollow Beach looking west, showing nature of Wellfleet plain outwash and diamicton lithofacies at detailed study section (after Stewart, 1988).

than those of the South Channel lobe (Oldale, 1976). During this differential retreat from west to east Glacial Lake Cape Cod was established in front of the Cape Cod Bay lobe (Larson, 1982). This section will focus on glaciogenic sediments in the Wellfleet Quadrangle that were deposited into Glacial Lake Cape Cod by the South Channel lobe, and also delimit part of its margin (Fig. 2).

## Geomorphology and stratigraphy of the Wellfleet Quadrangle

The Wellfleet Quadrangle of outer Cape Cod consists primarily of outwash of the Wellfleet plain (Fig. 2; Oldale, 1968; 1982). Radiocarbon dates suggest that the Wellfleet plain was deposited in late Wisconsinan time, about 20 to 27 Ka B.P. The upper surface of the Wellfleet plain slopes to the west, and its uncollapsed nature on the ocean side of Cape Cod suggests that the source was some distance offshore (Oldale, 1968). The Wellfleet plain is pitted by numerous kettle holes, attesting to the presence of stagnant ice during and after deposition.

The Wellfleet plain consists of two gross sedimentary lithofacies: (1) sand-and-gravel outwash deposited upon (2) silty, micaceous, stratified and massive diamictons containing sand and mud intrabeds. A deltaic origin is suggested by the geomorphology of the deposits, in particular their westward slope and the occurrence of deltaic foreset bedding and outwash cones (Stewart, 1988; Oldale, 1968). Exemplary exposures of both lithofacies occur at Newcomb Hollow Beach (Fig. 2). Sedimentary structures and particle fabrics in the diamicton lithofacies suggest that subaqueous debris flows and tractive currents were active during its deposition (Stewart, 1988). Abundant striations on clasts in the diamicton lithofacies suggest a basal origin; however, debris ultimately may have been released at higher levels from the glacier margin.

## Composition of the outwash and diamicton lithofacies

The outwash lithofacies can be separated into two subunits based on differences in clast lithologies (Oldale, 1968): the older and younger Wellfleet plain deposits (Fig. 3). With respect to younger Wellfleet plain deposits, older Wellfleet plain deposits contain lower percentages of granite and felsic volcanic rock, and higher percentages of mafic rock, quartzite, schist, and gray sedimentary rock (mainly sandstone).

The diamicton lithofacies consists mainly of granulometrically distinct massive and stratified diamictons. Massive diamictons contain an average of 8% sand (range 4-15%), 88% silt (range 80-92%) and 4% clay (range 2-5%), whereas stratified diamictons contain an average of 17% sand (range 9-29%), 74% silt (range 59-83%) and 9% clay (range 6-13%). The pebble suite of the diamicton lithofacies is distinguished by a predominance of phyllite and felsic volcanic rock, and a complete lack of quartzite and schist. Granite, mafic rock and vein quartz occur in lower quantities than in the Wellfleet plain deposits, which also contain slightly less red sedimentary rock than the diamicton lithofacies.

In keeping with the methodology of later sections of this paper (sections 4 and 5), which deal with the matrix of basal tills, geochemical analysis was performed only on matrix samples (finer than 0.063 mm fraction) of massive diamicton, whose parent material was probably basal debris (Stewart, 1988). The geochemical composition of massive diamicton matrix, in terms of four components (Fig. 4), is shown along with the

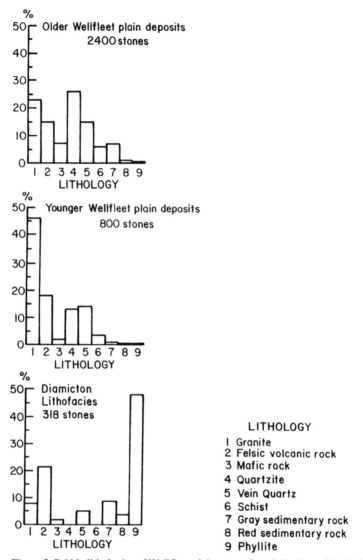

Figure 3. Pebble lithologies of Wellfleet plain outwash and diamicton lithofacies.

average matrix composition of Pleistocene till from Ontario (discussed below) and also the composition of Pettijohn's (1975) average slate. These other data are used to represent, respectively, two end members. The first end member depicts an unweathered feldspathic glacial rock flour in till from primary sources such as crystalline igneous and metamorphic rocks (see below). The second end member reflects debris weathered from primary sources and recycled as pelitic rocks and their metamorphosed equivalents in the Appalachian Geosyncline of New England (Fig. 2).

The geochemical data for the diamicton lithofacies fall about midway between the two end members. In terms of the normative mineralogical fields established for other tills

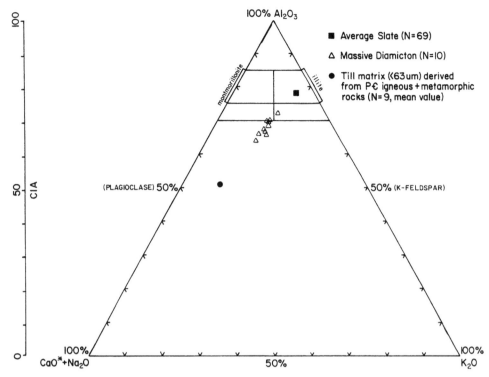

Figure 4. Geochemical composition of samples of massive diamicton matrix from Newcomb Hollow Beach. Also shown are compositions of the two end member components that are inferred to be the principal parent materials. Mineralogical fields and scale of CIA (Chemical Index of Alteration) as in Figure 14, and are shown here to facilitate later comparisons with other tills. Average slate composition after Pettijohn (1975); till compositions from Matachewan, Ontario (for further explanation, see text).

(Fig. 14), the massive diamictons can be viewed as a mixture of feldspathic rock flour and illitic (micaceous) debris. The normative mineralogy is reinforced by binocular micro-scope observations of abundant mica (10-25%) and feldspar in the coarse silt fractions of the massive diamicton matrix.

*Discussion*

The general path followed by the South Channel lobe (Fig. 2) is inferred from the following lines of evidence:

1. The geomorphology of the Wellfleet outwash plain and the sedimentary structures found in the outwash and diamicton lithofacies indicate a probable deltaic source that lay to the east of Cape Cod (Oldale, 1968; Stewart, 1988).

2. Pebble suites in the outwash and diamicton lithofacies have obvious similarities to rocks onshore, north and east of the study area. Red sedimentary rock (sandstone) and certain mafic rocks (diabase) are typical of Triassic basins (Oldale, 1976). Purple and red felsites are common in Silurian volcanic rocks of the Boston area (Howe, 1936; Kaktins,

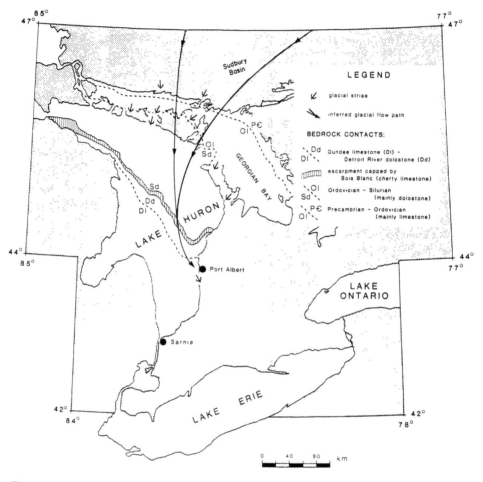

Figure 5. Location of Port Albert, Ontario, and approximate trajectory of the Huron lobe prior to deposition of the St. Joseph Till units. Legend indicates only principal bedrock types upglacier of Port Albert. Orientation of the glacial flow path is based on regional physiography, striae, glacial erratics, and till fabrics (Broster, 1982) (After Broster, 1986: Fig. 1; reproduced by permission of Canadian Journal of Earth Sciences).

1976). Gray quartzite clasts containing the Cambro-Ordovician brachiopod *Obolus* and found in the outwash and diamicton lithofacies are similar to clasts found in Pennsylvanian conglomerates that occur in Carboniferous basins of eastern New England (Walcott, 1898; Perkins, 1920). The abundant phyllite clasts in the diamicton lithofacies are comparable to phyllites that outcrop in eastern Massachusetts, southern New Hampshire and Maine. The overall affinity of the pebble suite in the diamicton lithofacies to known bedrock sources north and east of the study area implies that similar rocks must also occur in the Gulf of Maine. This contention is further supported by geophysical evidence and limited ocean bottom sampling (Fig. 2; Ballard and Uchupi, 1975).

3. The geochemistry of the diamicton matrix suggests contributions from two bedrock

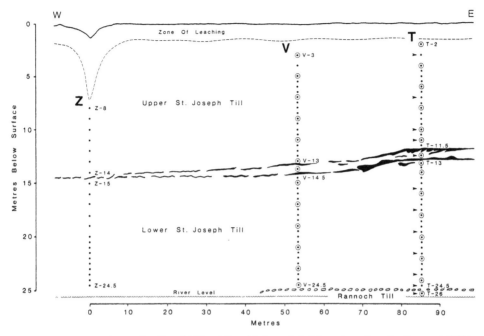

Figure 6. Location of sampling lines T, V and Z and informal division of stratigraphy. Rannoch Till is separated from the lower St. Joseph Till by a boulder horizon, and the dark lines midway up the section denote middle unit sands enclosing till. Sample locations are denoted by solid circles; samples analyzed for major elements, by solid circles within open circles; and locations of pebble collections, by arrows (after Broster, 1986: Fig. 2; reproduced by permission of Canadian Journal of Earth Sciences).

sources: one contributing a feldspathic rock flour and a second that provided mainly pelitic material. The rock flour component most likely derived from felsic igneous rocks and sedimentary or metamorphic rocks that could contribute silty felspathic detritus. The pelitic component most likely reflects contributions of kaolinite and illite (or mica) from argillaceous rocks that are common in the Appalachian geosynclinal belt (Fig. 2).

Active glaciers and analogous Pleistocene glacial margins that are dominated by fluvial systems and marked by heads-of-outwash (Fig. 1) are characterized by very complex processes and patterns of sediment dispersion (Evenson and Clinch, 1987). Because the geomorphology and sedimentology of the outwash and diamicton lithofacies suggests that their depositional environment was similar, no more than the previous general conclusions can be reached about the provenance of these deposits. The common occurrence of striated phyllite clasts in silty, micaceous diamicton suggests that the diamicton partly consists of phyllitic rocks that were comminuted in the basal debris zone of the South Channel lobe. The diamicton lithofacies lacks quartzite and schist, two pebble components that are common in the outwash lithofacies, which suggests that the outwash and diamicton lithofacies had debris sources of contrasting compositions.

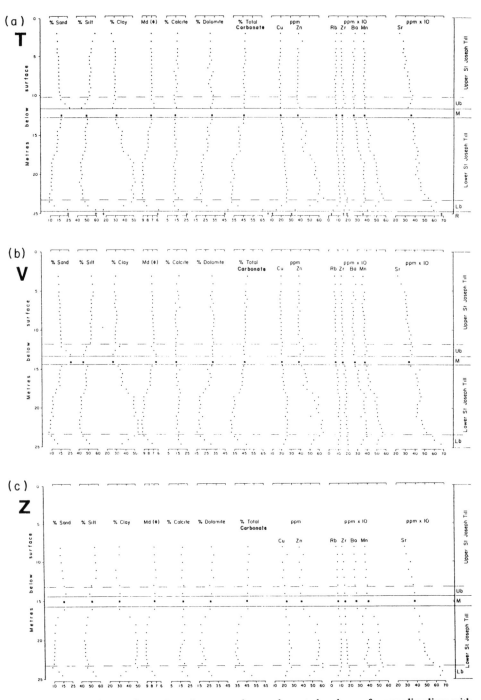

Figure 7. Variation in granulometric, carbonate, and trace-element abundances for sampling lines with depth. Solid lines denote separation in the middle St. Joseph Till unit (M) and Rannoch Till (R); broken lines denote basal subdivisions (Lb, Ub) of the lower and upper St. Joseph units (after Broster, 1986: Fig. 3; reproduced by permission of Canadian Journal of Earth Sciences).

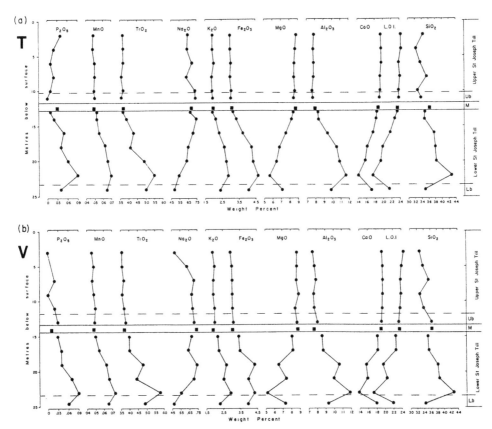

Figure 8. Variation of major elements with depth for sampling lines T and V. Elements are plotted from left to right in order of increasing aboundance within the units. Unit subdivisions are indicated as in Figure 7. (Total iron as $Fe_2O_3$; LOI, loss on ignition at 1000°C) (after Broster, 1986: Fig. 4; reproduced by permission of Canadian Journal of Earth Sciences).

### INTRA-TILL VARIABILITY: AN EXAMPLE FROM SOUTHERN ONTARIO, CANADA

To thoroughly understand the basis for compositional variations between till sheets (see below), it is first necessary to understand the nature and genesis of compositional variations within a till sheet. The following discussion will integrate pebble lithology, granulometry, carbonate mineralogy, major element and minor element geochemistry to characterize a till sheet in terms of its overall provenance, rather than individual aspects of the source materials.

### St. Albert Section

The study area is located near St. Albert, Ontario, and consists of a 25-metre-high section composed of the late Wisconsinan St. Joseph and Rannoch Tills (Fig. 5). Rannoch Till is only exposed near the river level and therefore was only minimally sampled. Three profiles were sampled in the St. Joseph Till, which here has been informally divided into the upper and lower units of the St. Joseph Till, which are separated by a thin till bed with intercalated sands (middle St. Joseph Till unit).

A total of 106 samples from three sampling lines (Fig. 6) was analyzed for granulometric composition (by sieve and hydrometer), carbonate mineralogy (by modified Chittick apparatus), and selected trace elements (by atomic absorption spectrophotometry). These results showed sufficient consistency between sections that sampling density for major elements could be reduced without a significant loss of information (Broster, 1986). A total of 28 samples from sampling lines T and V was analyzed using X-ray fluorescence techniques for the major elements Si, Ti, Al, Fe (total), Mn, Mg, Ca, K, Na, P, and loss-on-ignition, or LOI (mainly carbon dioxide and to a lesser extent water). Details of the analytical techniques have been discussed by Broster (1986).

### Compositional variation within the St. Joseph Till

Compositional variations within the subunits of the St. Joseph Till were subjected to a thorough statistical analysis to define the nature of compositional varability within and between sampling lines. Only the general statistical aspects of the study will be mentioned here, as the details have been discussed at length elsewhere (Broster, 1986). The data discussed are presented in Figures 7 and 8.

The basal zone of the lower St. Joseph Till demonstrates an upward decrease in the dilution of a clay-rich glacial debris load by entrainment of silt- and sand-rich Rannoch Till. Incorporation of Rannoch Till is further supported by similarities in clast lithologies (fig. 9). Relatively higher carbonate contents in Rannoch Till have also led to enrichment in CaO, MgO, LOI, Sr, percent calcite, percent dolomite and total carbonates in the basal zone of the lower St. Joseph Till. Low abundances of all other major and trace elements at the base suggest that they are associated with the carbonate-poor clay-size material that was already present within the glacier, prior to incorporation of Rannoch Till, and now constitutes the overlying main body.

In the main body of the lower St. Joseph Till statistical correlations with till textures suggested that the distribution of elements associated with silicate minerals (Si, Ti, Al, Fe, Mn, K, P) was controlled by the abundance of clay-sized material in the till. In contrast, the carbonate fraction was more strongly correlated with the quantity of silt, which is the typical modal size range for the terminal grade of carbonate minerals (Dreimanis and Vagners, 1971). The source of incorporated carbonate detritus is evident from the distribution of carbonate bedrock beneath the generalized path traversed by the Huron lobe (Fig. 5). In light of the general lack of shale clasts in the St. Joseph Till and the limited outcrop of shale in the Lake Huron basin, the source of the clay-sized material was probably incorporated older waterlaid sediments. This hypothesis is supported by (1) the nearby occurrence of waterlaid sediments below the St. Joseph Till, and (2) the known existence of a proglacial lake prior to the deposition of the St. Joseph Till.

The provenance of the main body of the lower St. Joseph Till is further established by upward trends in selected clast lithologies (Fig. 9). In the lower 6 metres of this unit, limestone alternates with dolostone as the most abundant lithology, whose variations are accompanied by a concomitant upward increase in the abundance of distal Precambrian clasts. The composition of the remainder of the section is more homogeneous than below.

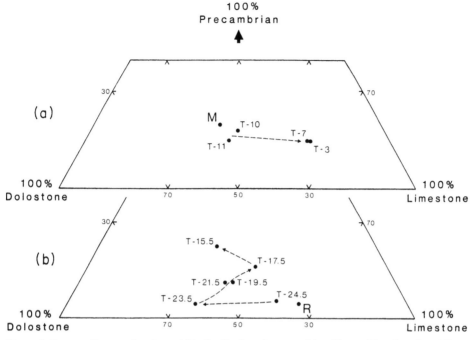

Figure 9. Ternary diagram showing pebble distributions for upper (a) and lower (b) units at Port Albert. Sample numbers decrease with elevation upward in the units (arrows); middle St. Joseph Till unit and Rannoch Till distributions are indicated by M and R, respectively (after Broster, 1986: Fig. 5; reproduced by permission of Canadian Journal of Earth Sciences).

Trace elements in this unit are strongly dependent on the quantity of clay present, implying a textural dependency (Fig. 7). This is not surprising considering the strong adsorptive capability of clay minerals (Knightly and Stewart, 1986). Inasmuch as field evidence in conjunction with the uniform trend of calcite with respect to variations in clay-sized material suggests that leaching has not been significant, it is likely that the trace and minor element concentrations are syngenetic and reflect the provenance of the debris load.

The middle unit of the St. Joseph till is chemically and lithologically similar to the top of the lower unit and the base of the upper unit. This similarity probably reflects incorporation of the lower unit prior to deposition of the middle unit.

The incorporation of till and intercalated sand of the middle unit into the glacier before deposition of the main body of the upper St. Joseph Till is expressed principally by clast lithologies and the enrichment of sand in the basal zone of the upper St. Joseph till (Figs. 7 and 9).

The main body of the upper St. Joseph till is distinguished by an overall homogeneity (Figs. 7 and 8). Statistical analysis revealed poor correlations between sand and clay, suggesting that grain size variations are controlled chiefly by changes in silt content. Clast lithologies (Fig. 9) are similar in the lower and upper parts of this zone. Limestone is about three times as abundant as dolostone, possibly due to preferential incorporation

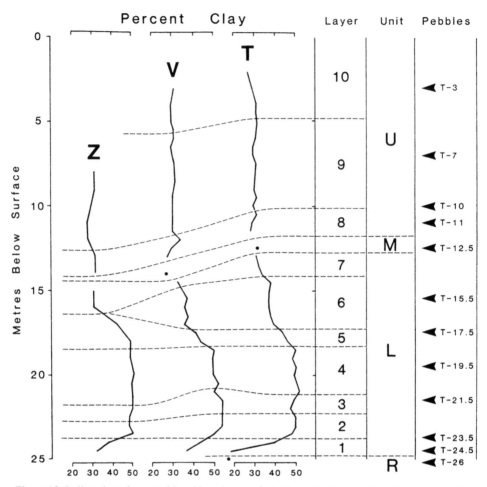

Figure 10. Delineation of compositional layering (1-10) by Q-mode cluster and multivariate discriminant function analysis and its relationship to informal unit divisions (R. Rannoch; L. lower St. Joseph; M, middle St. Joseph; U, upper St. Joseph) and pebble samples (e.g. T-19.5) (after Broster, 1986; Fig. 9; reproduced by permission of Canadian Journal of Earth Sciences).

of the softer limestone. This enrichment in limestone is not accompanied by an increase in calcite in the till matrix, which suggests that the transport distance of the limestone was too short to allow significant comminution to calcite. The uniformity of the percent total carbonates was also unexpected, since incorporation of bedrock by the Huron lobe was not isolated to a particular source, but encompassed a variety of lithologies, as evidenced by erratics of limestone, dolostone, Huronian rocks such as quartzite and diamictite, and indicators typically associated with the Sudbury Basin including pentlandite and Onaping tuff.

Major element concentrations throughout the main body of the upper St. Joseph Till are uniform, and in contrast to the lower unit, textural controls on major element distributions are lacking. The abundance and distribution of trace elements are also marked by their uniformity. The homogeneity of this unit is attributed to thorough mixing of different source materials during till formation.

Broster (1986) noted the presence of abnormal dolomite in the upper St. Joseph Till. Abnormal dolomite is a variety of dolomite that reacts in a similar way to calcite during gasometric carbonate analysis; moreover, the bedrock source of abnormal dolomite is limited to the Devonian Detroit River Group (Fig. 5; cf. Dreimanis, 1962). The occurrence of abnormal dolomite from a known area 8 to 45 km north of the Port Albert sections suggests that mixing and homogenization of the glacial load prior to till deposition was accomplished within this distance.

## Discussion

Elemental dispersion in the St. Joseph Till is believed to be syngenetic rather than epigenetic for the following reasons. Obvious post-depositional leaching of the study section was limited to a thin surface layer, which was not sampled (Fig. 6). Furthermore, the high clay content and dense texture of the tills has probably inhibited significant post-depositional alteration of the section by groundwater. This assertion is supported by the uniform trends in the upper unit of percent total carbonates and trace elements, all of which are susceptible to leaching (Ridler and Shilts, 1974). Although epigenetic dispersal of particularly mobile elements is possible, its effects are considered to be minor.

The dispersion patterns in the upper and lower units of the St. Joseph Till are inferred to be mainly a product of clastic dispersion. The mixing of source materials whose element correlations diverge would confuse correlation trends, whereas incorporation of materials with similar trends would reinforce correlations (Broster, 1986). Thus, homogenization of divergent source materials during incorporation and transport is believed to have obscured most definitive relationships in the upper unit. In the lower till, contributions from upglacier bedrock sources were gradually diluted by the entrainment of waterlaid sediments. Because upglacial units were incorporated sequentially, the relationship between upward-increasing carbonate detritus and decreasing silica-rich clay-sized sediment was reinforced.

The observed upward gradation from locally to distally derived material in the St. Joseph Till indicated that the units could be subdivided into stratified zones on the basis of composition. Compositional layering was defined on the basis of (1) principal components analysis of a data base consisting of variables reflecting sample depth, granulometric composition, trace element concentrations, and carbonate mineral content, followed by (2) Q-mode cluster analysis, which grouped samples on the basis of their bulk similarity. The statistical significance of these groupings was then tested by multivariate discriminant function analysis (Broster, 1986). The compositional stratification defined by statistical analyses is presented in Figure 10.

An accurate geological interpretation of individual layers is difficult because the layers consist of detritus from multiple sources. The basal layers of the upper and lower tills (layers 1 and 8) reflect incorporation of material from immediately below or a short distance upglacier. In the lower till, pebble lithologies of the lower four layers can be related to successive upglacier bedrock formations, but few relationships are apparent in the upper layers of either till (layers 5, 6 and 7, lower till; layers 9 and 10, upper till).

Chemically, each layer above layer 1 can be interpreted as a step in an overall transition upward from incorporation of silica-rich sources to incorporation of carbonate-rich

Figure 11. Simplified geological map of Iowa (after Anderson, 1983). Mudstones occur at various positions in all systems: 1-Cambrian; 2-Ordovician; 3-Silurian; 4-Devonian; 5-Mississippian; 6-Pennsylvanian; 7-Cretaceous.

sources (Figs. 7, 8). With regard to the lower till, two inferences can be made: (1) the downward increase in $SiO_2$ and $Al_2O_3$ is probably a result of increasing dilution of the till matrix by entrainment of local silica-rich clay, most likely waterlain sediments; and (2) the steady upward increase in abundance of matrix carbonates in the upper layers of the lower unit probably reflects the preferential comminution of intermediate-distance dolostones over distal, more resistant Precambrian silicate materials originating from north of Lake Huron. In the upper unit, $SiO_2$ also decreases upward inversely with CaO and MgO, but changes between layers are much less pronounced than in the lower unit.

The overall gradation of compositional layering demonstrated in the lower till favors a model of sequential glacial entrainment and retention of upglacier materials in higher ice positions. This stratification is in contrast to the general homogenization of the upper unit (Figs. 7, 8). Since the upper unit also contains erratics and material from the same upglacial sources as the lower unit, the greater mixing of the upper unit cannot be a function of increased transport distance but must be attributable to differences in glacial dynamics.

It is likely that no single mechanism is responsible for all differences between the upper and lower till units. The overall gradation in composition of the lower unit implies incorporation by basal accretion (Weertman, 1961; Boulton, 1970), whereas the homogenization of the upper unit is believed to be the result of extensive shearing caused by greater resistance along the glacier bed, resulting in shear stacking and confusion of the original debris sequence (Virkkala, 1952; Moran, 1971). Mixing and homogenization of material deposited in the upper unit was probably accomplished within 8 to 45 km of glacial transport, which underscores the rapidity of the process.

The glacial theory established by previous workers (Virkkala, 1952; Weertman, 1961; Boulton, 1970; Moran, 1971; Kemmis, 1981) implies that entrainment processes promote development of intra-till compositional stratification, although englacial debris tends to become more homogeneous with transport. Factors contributing to homogenization during transport are those conditions supporting increased thrusting and shearing within glacial ice. Thrusting and shearing are interpreted to have been less extensive during erosion, transport and deposition of material constituting the lower St. Joseph till, which allowed preservation of compositional layering.

INTER-TILL VARIABILITY IN TILLS FROM IOWA, U.S.A.

*Geology of Iowa and related areas*

The bedrock of Iowa is predominantly Paleozoic and Mesozoic (Cretaceous) clastic and carbonate bedrock (Fig. 11). As in most states in the terminal zone of the Laurentide ice sheet, the bedrock of Iowa is almost entirely mantled by Pleistocene drift of varying composition, age and thickness. Multiple Pleistocene tills record at least six major episodes of glaciation in eastern Iowa and eight in western Iowa-eastern Nebraska (Figs. 12, 13). Non-glacial intervals are characterized by paleosols and loess deposits (Knightly and Stewart, 1986).

The sand fraction of till matrix in Iowa has been studied extensively through a variety of petrographic techniques, whereas the silt-plus-clay fraction is typically analyzed only for its calcite and dolomite content (Ruhe, 1969; Boellstorff, 1978; Hallberg, 1980; Kemmis et al., 1981). These studies have demonstrated a rather homogeneous intra-till character. In light of this extensive body of previous work, the following discussion will focus on the nature of the silicate phases in the silt-plus-clay fraction of till matrix, which commonly constitutes 50-75% of most till samples in Iowa.

*Methods*

Till sheets in Iowa typically are composed of basal tills, whereas supraglacial deposits that contain mainly till-like sediments are most commonly found in end moraines (Hallberg, 1980; Kemmis et al., 1981). Chemical analyses of basal tills, supraglacial tills and till-like sediments, and bedrock were reduced to normative mineral compositions following the methods outlined by Nesbitt and Young (1982).

All chemical analyses of Iowa bedrock are of mudstones. Other analyses included for comparative purposes include Archean Morton Gneiss from Minnesota and Proterozoic Gowganda Formation diamictites from the Matachewan area of northern Ontario, and nineteen analyses of Pennsylvanian mudstones from Illinois. These data and respective analytical techniques are discussed in the following sources: Weems (1904) and Beyer and Williams (1904) (Iowa); White (1959) (Illinois); Goldich (1938) (Minnesota), and Knightly (1987) (Ontario).

Feldspars are the most abundant of the labile minerals in the Earth's upper crust (Wedepohl, 1949). Calcium, sodium and potassium are removed from feldspars by

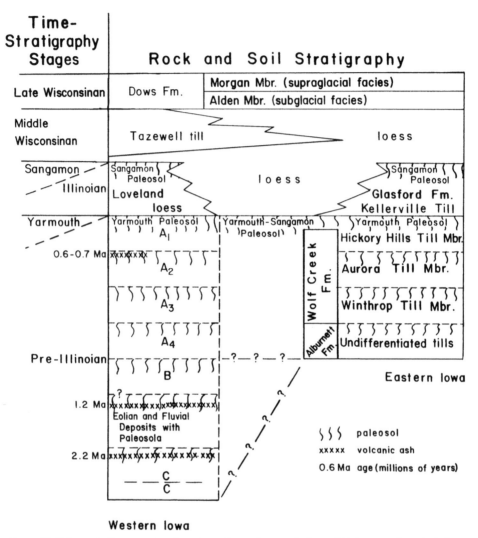

Figure 12. Schematic diagram of stratigraphy and tentative correlations of Pleistocene deposits in Iowa (after Knightly and Stewart, 1986). Individual pre-Illinoian tills in western Iowa are designated by capital letters (A, B, C). Position of paleosols is relative.

chemical weathering (Birkeland, 1984), therefore, the proportion of residually enriched alumina to labile element oxides will increase in the weathered product (Kronberg and Nesbitt, 1981; Nesbitt and Young, 1982). Nesbitt and Young (1982) found that the degree of weathering could be expressed by a chemical index of alteration (CIA), using oxide mole proportions:

CIA = $(Al_2O_3/(Al_2O_3 + CaO^* + Na_2O + K_2O)) \times 100$ (Equation 1) where $CaO^*$ is the molar quantity of CaO in silicate phases. The end members in this calculation [A1 = CIA = (mole % $Al_2O_3$), Ca + Na = (mole % CaO + $Na_2O$) and K = (mole % $K_2O$); each with

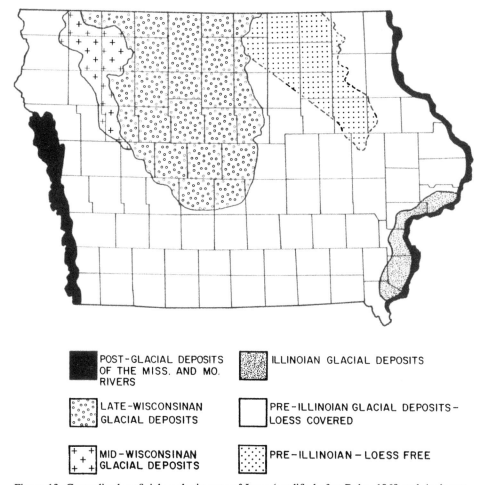

POST-GLACIAL DEPOSITS
OF THE MISS. AND MO.
RIVERS

ILLINOIAN GLACIAL DEPOSITS

LATE-WISCONSINAN
GLACIAL DEPOSITS

PRE-ILLINOIAN GLACIAL DEPOSITS–
LOESS COVERED

MID-WISCONSINAN
GLACIAL DEPOSITS

PRE-ILLINOIAN – LOESS FREE

Figure 13. Generalized surficial geologic map of Iowa (modified after Ruhe, 1969 and Anderson, 1983). Illinoian deposits are also largely loess-covered, as are middle-Wisconsinan deposits, although more sporadically. Late-Wisconsinan deposits have little or no loess mantle. Reproduced by permission of the National Water Well Association.

respect to the denominator in equation 1] were used to calculate normative chemical compositions for unaltered feldspars and the three typical alteration products kaolinite, illite and montmorillonite (Fig. 14; cf. Garrels and Mackenzie, 1971, Appendix 1). Note that only the upper part (CIA > 50) of the ternary diagram is shown in Figs. 14, 15 and 16. Plagioclase (50% albite, 50% anorthite) has a CIA of 50 and plots on the $(CaO^* + Na_2O) -$ $(Al_2O_3)$ boundary; similarly, potassium feldspar plots on the opposite $(K_2O - Al_2O_3)$ boundary, also with a CIA of 50. Kaolinite, which is essentially stripped of labile elements, has a CIA of nearly 100 and plots at or near the $Al_2O_3$ apex of the triangular diagram. Montmorillonite plots along the $(CaO^* + Na_2O) - (Al_2O_3)$ boundary with a CIA of 75-85. Illite (CIA = 75-85) plots in a corresponding position on the $(K_2O - Al_2O_3)$ boundary. In general, tills and bedrock will exhibit an increasing CIA (i.e.

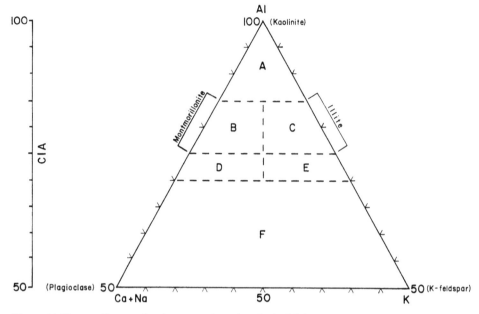

Figure 14. Ternary diagram showing normative mineralogical fields and end-member compositions.

proportion of $Al_2O_3$) in response to (1) feldspar alteration (removal of Ca, Na, and K) and (2) admixture or incorporation of clay minerals.

*Results*

For ease of presentation and interpretation, the upper half of a ternary diagram (Fig. 14) has been divided into six fields. Whereas the actual matrix of till is a mixture of various minerals, the specific results presented below should be viewed in terms of general compositions. Figure 14 permits a rapid assessment of the matrix mineralogy of a given sample and in particular its clay mineralogy. Field A consists of predominantly kaolinitic clay, whereas fields B and C are composed of a montmorillonite component (B) and an illite component (C). Fields D and E are analogous to, respectively, fields B and C; however, they contain less of the clay mineral component of each field and are enriched in a silty 'rock flour' component consisting of plagioclase feldspar (field D) or potassium feldspar (field E). Field F consists of silt-sized, unaltered feldspar.

   Geochemical analyses of bedrock are presented in Figure 15. Precambrian bedrock samples consist mainly of unaltered plagioclase feldspar and lesser amounts of potassium feldspar (field F). Most of the feldspar found in Iowa tills derives from similar Precambrian bedrock in the upper midwest (Minnesota, Wisconsin and Michigan: Kemmis et al., 1981). Feldspar liberated from Precambrian rocks during Phanerozoic time was repeatedly weathered and recycled in a variety of clastic sedimentary rocks; in Iowa and adjacent Illinois these rocks are chiefly mudstones (Figs. 11, 15; Anderson, 1983). Ordovician mudstones in Iowa are similar to Precambrian rocks in that they also contain mostly fine-grained plagioclase feldspar (field F). Silurian mudstones are generally more potassic, plotting mainly in field C. Devonian mudstones consist of rock flour of a variable composition (fields D, F). Mississippian mudstones are illite-rich (fields C and E). Pennsylvanian mudstones contain kaolinite, illite and montmorillonite and plot in fields A, B, and C. The Iowa samples contain more of the montmorillonite component, whereas the Illinois samples are more illitic. Cretaceous mudstones are similar to their Pennsylvanian counterparts, plotting mostly in fields A and B.

The composition of Iowa tills is shown in Figure 16. The matrix of till derived from Archean bedrock from Matachewan, Ontario, Canada, shown for comparison, is predominantly a plagioclase-rich rock flour, consisting of chiefly silt-sized material with less than 1-2% clay (Stewart and van Hees, 1983). Wisconsinan till from Iowa also has a silt-rich matrix; however, the clay fraction is more abundant (15%) and contains montmorillonite and illite (Kemmis et al., 1981, pp. 27, 43). For this reason, the data plot near the upper boundary of field F, and equidistant between fields D and E. Younger supraglacial tills (Morgan Member) are slightly less aluminous than older basal tills (Alden Member and Mid-Wisconsinan Tazewell till).

Illinoian tills plot among fields B through F. These samples are variably enriched in aluminous material as compared to Wisconsinan tills. This is to be expected inasmuch as the average clay content of Illinoian tills (28%) is nearly twice that of Wisconsinan tills (Wickham, 1980, table 1).

The clay content of pre-Illinoian tills averages 23-25% (Hallberg, 1980, p. 16, 79); moreover, their compositions (Fig. 16) are comparatively enriched in aluminous minerals with respect to Wisconsinan and Illinoian tills, even though the average clay content of the latter is slightly higher. Most samples of pre-Illinoian till plot in fields C and E, and to a lesser extent in fields D and F, suggesting a mineralogy of, in decreasing abundance, illite, montmorillonite and kaolinite.

*Discussion: the relationship of till to bedrock in Iowa*

The Iowa landscape was subjected to extensive weathering and fluvial erosion during the Tertiary Period (Anderson, 1983), resulting in a rolling topography with many broad valleys which partly guided the movement of the oldest (pre-Illinoian) glaciers entering the state. Tills deposited during the oldest advances hence bear a strong geochemical imprint from local bedrock of high CIA (Fig. 17). Pennsylvanian and Cretaceous rocks comprise about two-thirds of the subcrop beneath the Pleistocene cover in Iowa, so it is not surprising that their influence is reflected strongly by pre-Illinoian tills. Deposits of successively younger glaciers progressively buried the weathered preglacial Iowa landscape and older till sheets, leaving less bedrock available for direct incorporation into till. At the time the Des Moines lobe advanced into Iowa in late Wisconsinan time, probably less than 10% of its substrate was bedrock (Palmquist and Bible, 1974; Kemmis et al., 1981).

The observed matrix compositions of Illinoian and older tills in Iowa may reflect compositional variations in the bedrock eroded, whose influence diminished during burial by successive tills. It has been shown elsewhere (Ruhe, 1969; Kemmis et al., 1981) that following pre-Illinoian and Illinoian glaciations, later glacial advances incorporated relatively little of older till. The data presented above support these contentions. The matrix debris load of successively younger glaciers that deposited the mid-Wisconsinan Tazewell till and late Wisconsinan Dows Formation was apparently dominated by

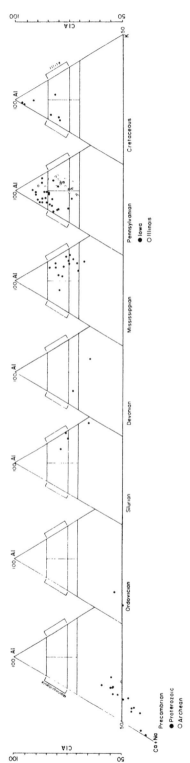

Figure 15. Compositional variation of Precambrian rocks and Phanerozoic mudstones in the study area. Boundaries and end members correspond to Figure 14. Each circle represents one analysis.

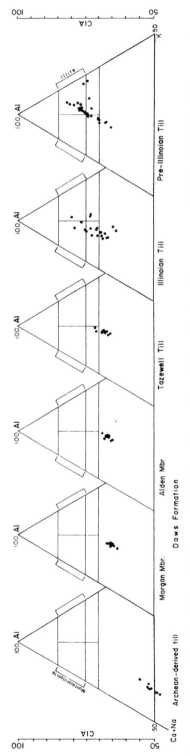

Figure 16. Compositional variation of Pleistocene tills in the study area. Boundaries and end members correspond to Figure 14. Each circle represents one analysis.

low-CIA rock flour mainly derived from Precambrian rocks in the upper midwest. This material ultimately became the principal constituent of Wisconsinan tills without as significant dilution by local high-CIA material as in older tills (Fig. 17).

The slight geochemical contrast between the supraglacial and subglacial units of the late Wisconsinan Dows Formation may reflect two genetic factors. Firstly, supraglacial debris typically has a more distal provenance than basal debris (Boulton, 1970; Dreimanis, 1976). Secondly, supraglacial debris is often reworked by sediment gravity flows, which may winnow the fine fraction, and in doing so remove argillaceous material of high CIA. Both hypotheses may be applicable to the Morgan Member. The abundant evidence of resedimentation in this unit, plus its sandier texture relative to the basal Alden Member (Kemmis et al., 1981) suggest that the second hypothesis may have more validity than the first. Kemmis (1981) suggested that the homogeneity of Alden Member basal till was due to extensive regelation at the base of an ice sheet, which caused repeated episodes of incorporation, transport and deposition of debris.

It is interesting to note that the composition of till matrix derived from Archean crystalline rock at Matachewan, Ontario is similar to that of till derived from crystalline bedrock in the Precambrian shield of Finland (Taipale et al., 1986). In both cases, plagioclase is considerably more abundant than potassium feldspar in the fine fraction of the till matrix, which implies that the gross mineralogy of debris eroded from crystalline sources is compositionally biased toward plagioclase. It can therefore be predicted that the plagioclase-rich rock flour component will be a principal constituent of till matrix in the marginal area of the Fennoscandian ice sheet in central Europe, whose bedrock geology is also generally similar to the southern Great Lakes region of the United States.

The lack of significant incorporation of older drift sediments during later glacial events probably reflects the readily deformable character of the older drift. Such sediments promote high subglacial pore water pressures, which limit the efficacy of basal erosion and incorporation (Boulton and Jones, 1979). In the shield terrain of the upper midwestern USA (Minnesota, Michigan) and adjacent parts of Canada, local bedrock relief is considerably greater and bedrock is generally harder than farther south, which favors repeated glacial erosion of fresh, geochemically immature bedrock. The erosion products persisted to become the predominant components of till matrix in the marginal areas of the Laurentide ice sheet.

CONCLUSIONS

The silt-plus-clay matrix fraction of till is particularly useful for provenance studies for two reasons: (1) it typically contains the comminution products of glacial erosion which constitute the bulk of material in most tills, and (2) it is also sensitive to compositional changes that occur due to incorporation of fine-grained sources such as lake sediment or shale.

End moraine depositional environments are complex because of resedimentation processes that may commonly affect tills and associated glaciofluvial sediments. For this reason the transport history of tills, till-like sediments, outwash and any indicator clasts will be obscured. The example of diamicton compositions from Cape Cod illustrates how unweathered (rock flour) and weathered (pelitic) bedrock sources contributed to the debris load of the South Channel lobe. The head-of-outwash environment studied in the

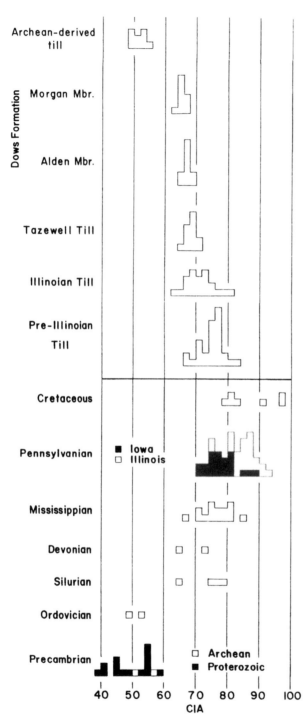

Figure 17. Summary diagram showing variation of CIA (chemical index of alteration) among samples of bedrock and various Pleistocene tills. Each square represents one analysis.

Wellfleet area and others like it can be used to obtain a first approximation of the general provenance of part of an ice sheet. Trace metal anomalies or clasts bearing ore minerals should be followed up by regional provenance studies of till sheets behind the head-of-outwash or end moraine of interest.

Generally in marginal areas, till sheets should become more homogeneous in areas characterized by abundant thrusting and shearing within the ice mass, otherwise, till sheets will likely be compositionally layered. Throughout the till sheet, the occurrence of indicator clasts will be biased toward preservation of the more resistant upglacial lithologies, whereas the matrix properties will be biased toward both the less resistant bedrock lithologies and minerals, and the composition of any incorporated unconsolidated sediments. The significance of indicator clasts in stacked till sheets will depend on how well the host till unit can be traced on a regional basis. If this stage of an exploration program is successful, then further work should proceed to determine the occurrence of the indicator on an intra-till basis.

ACKNOWLEDGEMENTS

This study has been funded by grants from the Iowa Science Foundation, the Iowa Geological Survey, the Iowa State University Achievement Foundation, and Pamour Porcupine Mines Ltd. (Timmins, Ontario) to Stewart. Additional support for publication was received from Natural Science and Engineering Research Council grant 9697 to Broster. We thank the anonymous reviewers for improving the clarity of the manuscript.

REFERENCES

Anderson, R.C. 1957. Pebble and sand lithology of the major Wisconsin glacial lobes of the central lowland. Geological Society of America Bulletin 68: 1415-1450.

Anderson, W.I. 1983. Geology of Iowa. 268 pp. Ames. IA: Iowa State University Press.

Ballard, R.D. and E. Uchupi 1975. Triassic rift structure in Gulf of Maine. American Association of Petroleum Geologists Bulletin 59: 1041-1072.

Beyer, S.W. and I.A. Williams 1904. The geology of clays. Iowa Geological Survey Report 14: 381-554.

Birkeland, P.W. 1984. Soils and Geomorphology. 372 pp. New York: Oxford University Press.

Boellstorf, J.D. 1978. Procedures for the analysis of pebble lithology, heavy minerals, light minerals, and matrix calcite-dolomite of tills. In G.R. Hallberg (ed.), Standard procedures for evaluation of Quaternary materials in Iowa: 31-60. Iowa City, IA: Iowa Geological Survey.

Boulton, G.S. 1970. On the origin and transport of englacial debris in Svalbard glaciers. Journal of Glaciology 9: 213-229.

Boulton, G.S. 1972. Modern Arctic glaciers as depositional models for former ice sheets. Journal of the Geological Society of London 128: 361-393.

Boulton, G.S. and A.S. Jones 1979. Stability of temperate ice caps and ice sheets resting on beds of deformable sediment. Journal of Glaciology 24: 29-44.

Broster, B.E. 1982. Compositional variations in the St. Joseph till units of the Goderich area, Ontario. Ph. D. thesis, London, Ontario: The University of Western Ontario.

Broster, B.E. 1986. Till variability and compositional stratification: examples from the Port Huron lobe. Canadian Journal of Earth Sciences 23: 1823-1841.

Dreimanis, A. 1962. Quantitative gasometric determination of calcite and dolomite by using Chittick apparatus. Journal of Sedimentary Petrology 32: 520-529.

Dreimanis, A. 1976. Tills: Their origin and properties. In R.F. Legget (ed.), Royal Society of Canada Special Publication 12: 11-49. Ottawa, Ontario: Royal Society of Canada.

Dreimanis, A. and U.J. Vagners 1971. Bimodal distribution of rock and mineral fragments in basal tills. In R.P. Goldthwait (ed.), Till – A Symposium: 237-250. Columbus, OH: Ohio State University Press.

Evenson, E.B. and J.M. Clinch 1987. Debris transport mechanisms at active alpine glacier margins: Alaskan case studies. Geological Survey of Finland, Special Paper 3: 111-136.

Garrels, R.M. and F.T. MacKenzie, 1971. Evolution of Sedimentary Rocks. 397 pp. New York: W.W. Norton.

Goldich, S.S. 1938. A study of rock weathering. Journal of Geology 46: 17-58.

Hallberg, G.R. 1980. Pleistocene stratigraphy in east-central Iowa. Iowa Geological Survey Technical Information Series 10, 168 pp.

Howe, O.H. 1936. The Hingham red felsite boulder train. Science 84: 394-396.

Kaktins, U. 1976. Stratigraphy and petrography of the volcanic flows of the Blue Hills area, Massachusetts. Geological Society of America Memoir 146: 125-142.

Kemmis, T.J. 1981. Importance of the regelation process to certain properties of basal tills deposited by the Laurentide ice sheet in Iowa and Illinois, U.S.A., Annals of Glaciology 2: 147-152.

Kemmis, T.J., G.R. Hallberg and A.J. Lutenegger 1981. Depositional environments of glacial sediments and landforms on the Des Moines lobe, Iowa. Iowa Geological Survey Guidebook Series 6, 132 pp.

Knightly, J.P. 1987. The stratigraphy and sedimentology of the Precambrian Gowganda Formation near Matachewan, Ontario, Canada. Unpublished M.S. thesis, Ames, IA: Department of Earth Sciences, Iowa State University.

Knightly, J.P. and R.A. Stewart 1986. Some geochemical aspects of Iowa Pleistocene sediments and their bedrock sources: Implications for the attenuation of agricultural and other ground water contaminants. Agricultural Impacts on Ground Water: 580-596. Dublin, OH: National Water Well Association.

Kronberg, B.I. and H.W. Nesbitt 1981. Quantification of weathering, soil geochemistry and soil fertility. Journal of Soil Science 32: 453-459.

Larson, G.J. 1982. Nonsynchronous retreat of ice lobes from southeastern Massachusetts. In G.J. Larson and B.D. Stone (eds.), Late Wisconsinan Glaciation of New England: 101-114. Dubuque, IA: Kendall/Hunt.

Lawson, D.E. 1979. Sedimentological analysis of the western terminus region of the Matanuska Glacier, Alaska. Cold Regions Research and Engineering Laboratory (CRREL) Report 79-9, Hanover, NH: CRREL.

Mather, K.F., R.P. Goldthwait and L.R. Theismeyer 1942. Pleistocene geology of western Cape Cod, Massachusetts. Geological Society of America Bulletin 53: 1127-1174.

Mickelson, D.M., L. Clayton, D.S. Fullerton and H.W. Borns Jr. 1983. The late Wisconsin glacial record of the Laurentide ice sheet in the United States. In S.G. Porter (ed.), Late-Quaternary Environments of the United States, Vol. I, The Late Pleistocene: 3-37. Minneapolis, Minnesota: University of Minnesota Press.

Moran, S.R. 1971. Glaciotectonic structures in drift. In R.P. Goldthwait (ed.), Till – A Symposium: 127-148. Columbus, OH: Ohio State University Press.

Nesbitt, H.W. and G.M. Young 1982. Early Proterozoic climates and plate motions inferred from major element chemistry of lutites. Nature 299: 715-717.

Oldale, R.N. 1968. Geologic map of the Wellfleet quadrangle. Barnstable County, Cape Cod, Massachusetts. U.S. Geological Survey Geologic Quadrangle Map GQ-750.

Oldale, R.N. 1976. Notes on the generalized geologic map of Cape Cod. U.S. Geological Survey Open-File Report 76-765, 23 pp.

Oldale, R.N. 1982. Pleistocene stratigraphy of Nantucket, Martha's Vineyard, the Elizabeth Islands, and Cape Cod, Massachusetts. In G.J. Larson and B.D. Stone (eds.), Late Wisconsinan Glaciation of New England, pp. 134. Dubuque, IA: Kendall/Hunt.

Palmquist, R.C. and G. Bible 1974. Bedrock topography beneath the Des Moines lobe drift sheet, north-central Iowa. Proceedings Iowa Academy of Sciences 81: 164-170.

Perkins, E.H. 1920. The origin of the Dighton conglomerate of the Narragansett basin of Massachusetts and Rhode Island. American Journal of Science, 4th series, 49: 61-75.

Pettijohn, F.J. 1975. Sedimentary Rocks. 3rd. ed., 274 pp., New York: Harper and Row.

Ridler, R.H. and W.W. Shilts 1974. Exploration for Archean polymetallic sulphide deposits in permafrost terrains: An integrated geological/geochemical technique; Kaminak Lake area, District of Keewatin. Geological Survey of Canada Paper 73-34, 33 pp.

Ruhe, R.V. 1969. Quaternary Landscapes in Iowa. 255 pp. Ames, IA: Iowa State University Press.

Schafer, J.P. 1980. The last ice sheet in Rhode Island. Geological Society of America, Abstracts with Programs 12: 80.

Stewart, R.A. 1988. Stratigraphy and provenance of diamictons and outwash deposited by the South Channel lobe in Eastham and Wellfleet, Cape Cod, Massachusetts. In J. Brigham-Grette (ed.), Field Trip Guidebook for AMQUA 1988: 61-100. Contribution No. 63 to the Department of Geology and Geography, University of Massachusetts: Amherst, Massachusetts.

Stewart, R.A. and E.H. van Hees 1983. Evaluation of past-producing gold mine properties by drift prospecting: An example from Matachewan, Ontario, Canada. In E.B. Evenson, Ch.Schlüchter and J. Rabassa (eds.), Tills and Related Deposits: 179-193. Rotterdam: Balkema.

Taipale, K., R. Nevalainen and M. Saarnisto 1986. Silicate analyses and normative compositions of the fine fraction of till: Examples from eastern Finland. Journal of Sedimentary Petrology 56: 370-379.

Virkkala, K. 1952. On the bed structure of till in eastern Finland. Bulletin de la Commission geologique de la Finlande 157: 97-109.

Walcott, C.D. 1898. Note on the brachiopod fauna of the quartzite pebbles of the Carboniferous conglomerates of the Narragansett Basin, Rhode Island. American Journal of Science, 4th series, 6: 327-328.

Wedepohl, K.H. 1969. Handbook of Geochemistry 1. Berlin: Springer-Verlag.

Weems, J.B. 1904. The chemistry of clays. Iowa Geological Survey Report 14: 321-345.

Weertman, J. 1961. Mechanism for the formation of inner moraines found near the edges of cold ice caps and ice sheets. Journal of Glaciology 3: 965-978.

White, W.A. 1959. Chemical and spectrochemical analyses of Illinois clay materials. Illinois State Geological Survey Circular 282, 55 pp.

Wickham, J.T. 1980. Status of the Kellerville Till Member in Western Illinois. Iowa Geological Survey Technical Information Series 11: 151-180.

Woodworth, J.B. and E. Wigglesworth 1934. Geography and geology of the region including Cape Cod, the Elizabeth Islands, Nantucket, Martha's Vineyard, No Man's Land, and Block Island. Harvard College Museum of Comparative Zoology Memoir 52, 322 pp.

# CHAPTER 9

# Lithology and transport distance of glaciofluvial material

MARIANNE LILLIESKÖLD

*Department of Quaternary Research, University of Stockholm, Sweden*

## INTRODUCTION

Research into the lithology and transport distance of glaciofluvially derived material has a long tradition. In the late 19th century glacial erratics were used as indicators of the maximum extent of the Pleistocene glaciers, and it was found that lithological variations and grain shapes were related to the transport mode and distance. In a study on glacial tills and glaciofluvial material transported at least 200 km, von Post (1855) came to the conclusion that the angular material was transported by drift ice and dropped instantaneously. The form and roundness of the clasts led him to conclude that the esker had been built up by ocean waves and currents.

In one of the first papers on differences in lithology, Gumaelius (1871) pointed out that the close relationship between tills and the underlying bedrock was not seen in glaciofluvial deposits. Later Gumaelius (1885) also stressed the importance of proper sampling from sections, in contrast to von Post, who picked boulders from the gravel pit floor.

Much research has been carried out in glaciated terrain since that time (Stone 1899, Hellaakoski 1930, G. Lundqvist and Hjelmqvist 1941, Trefethen and Trefethen 1944, Okko 1945, Virkkala 1958, Gillberg 1968, Lee 1965, Flint 1971, Shilts 1973, etc.), and most authors have achieved similar results, indicating rather short glacial meltwater transport, while only one investigation into a long esker in Central Sweden has shown extremely long transport distances. In their detailed provenance investigations at active glacier margins in Alaska, Evenson and Clinch (1987) demonstrate the importance of fluvial systems in alpine environments. The large quantities of material deposited by fluvial processes (90%) indicate that this activity is the dominant debris transport mechanism in this environment and not glacial transport. The purpose of this paper is to give an introduction to the factors influencing the lithological composition of esker material. Some cases illustrating transport distances in glaciofluvial systems are reviewed, and the applicability of this material to indicator tracing is discussed in the light of two research areas in Sweden.

## THEORY

### Glacial meltwater erosion

The origin of glaciofluvial material is either the bedrock, till or englacial debris. In the first two cases erosive forces cause extraction of the material into the meltwater system. Processes such as plucking and abrasion initiate subglacial erosion due to shear stresses exerted by the overriding ice, and then regelation processes (Boulton 1979, Boulton et al. 1974, Hallet 1979, Röthlisberger and Iken 1981) incorporate the loosened bedrock material into the sole of the glacier.

Only recently has the role of meltwater been considered relevant in glacial erosion (Drewry 1986). Many factors interact in the meltwater erosion process. Due to the very low temperatures, the viscosity of the water can be high and together with fragile ice or heavy loaded high velocity flow, meltwater can exert an extremely abrasive action. Erosion by impact fracturing and abrasion by suspended and bed load transport are also significant erosional processes.

There are no major differences between subglacial and open channels with respect to erosion phases, but the bedrock properties together with hydrological features are the main controlling factors in these processes. Abrasion is the most important mechanism causing erosion at lower velocities, while cavitation is important at velocities higher than 12 m/s.

## Particles during transport

Assuming that glaciofluvial material emanates from till or directly from debris-rich ice, the particles already have an initial form. Reworking begins as this material enters the glaciofluvial system. Particles moved along the bed are exposed to shear stress in traction and point loading during saltation. This leads to propagation along inter-crystal planes within the particles, resulting in smaller monominerals and lithic fragments (Slatt & Eyles 1981). Abrasion by impinging suspended particles during flow is another important mechanism acting during transport.

## Deposition

Sediments transported by glacial meltwater are deposited principally under conditions of diminishing or recessional flow. The mode of deposition controls the lithology, the stratigraphy and the facies assemblages. There are scarcely any differences between bed forms and structures formed in glacial meltwater tunnels or open conduits (Bannerjee and McDonald 1975).

The lithological variation in different beds usually reflects the grain-size distribution within the internal structures, though there is some indication of differences in bedrock content between separate beds within the same size fraction. In subglacial glaciofluvial deposits, the percentage of the local bedrock usually rises due to continual unloading of material, while in the englacial system the longest-transported material is entrained.

In proglacial environments, material released from the meltwater conduit and carried into a braided system is transported still further downstream from its source. Shifting flow regimes due to overloaded streams cause grain-size separation which in turn has an effect upon the petrographical composition of deposits with respect of long-range transported material.

The transported particles becomes reduced in grain size in the above processes, and there is also a clear trend for the median grain size to diminish with distance from glacier terminus in outwash streams (Slatt and Eyles 1981, Rice 1982, Drewry 1986).

## Grain shape

The concept of grain shape covers a combination of three elements: shape, roundness and sphericity. A fragment of a certain rock type which has recently been plucked or crushed will have a similar grain shape whatever the size, due to the influence of its provenance. The processes acting upon the particles depend on the mode and rate of transport, and will altogether give the particle its final roundness with smoothened edges and more or less sphericity. Many attempts have been made to express the properties of the complex shape. Some of the most useful measurements are the'roundness scale' (Powers 1953) and'form classes' (Zingg 1935, see also Hirvas & Nenonen, this volume). The present author is testing a unique instrumentation for automatic image analyses of each individual grain within a sample of glacial sands and gravels (0.125–10 mm), including measurement of the third axis (Lilliesköld 1990).

## METHODS

### Sampling technique

Boulders and cobbles are taken out from the section wall and the bedrock lithology is determined in the field. Stone counts at different levels are used to describe the composition of the whole section.

The finer material is sampled in clearly separated layers. Only grains deposited under the same hydrodynamic conditions should be sampled, in order to obtain good correlations. Regional variations are shown by differences in the mean values for the sections.

In the laboratory the samples can be split and fractionated. Grain-size fractions of 2–5.6 mm and 8–16 mm were analysed in the Badelunda and Indal investigations. Special care must be taken when sampling easily disintegrating material to avoid under-representation in coarser fractions or enrichment of finer sizes. Hellaakoski (1930) estimated this error to be 2–3% for rapakivi granite.

The sampling interval may have to be varied on account of the comminution rates for different bedrock lithologies, the excavations available or interruptions in the course of the esker.

The number of grains counted varies from 200 to 500. In homogenous glaciofluvial material 100 + 100 show no differences, while in the unsorted material in tills 250–500 grains have to be counted (J. Lundqvist 1952, Persson 1975).

### Conducting and presentation of counts

The transport distances of glaciofluvial material can be determined by choosing lithological tracers which are easy to recognize, even in finer grain sizes. Some problems may arise if the parent rock is coarse-grained, since the gravel will then be monomineralic in composition and its relation to bedrock source is more difficult to determine.

A binocular microscope is usually used to analyse sand and gravel fractions, but thin section examination, X-ray diffraction and heavy mineral analyses are often included in special investigations of the fine fractions of glaciofluvial material.

The results of the frequency counts can be presented in many ways. In the graphic method, which is the most common, each bedrock type is expressed as a percentage of the

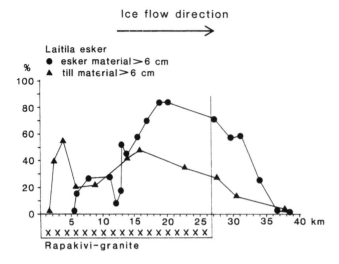

Figure 1. The relation between glacial and glaciofluvial transport in the area of the Laitila esker, Finland, according to the data of Hellaakoski (1930). The indicator rock (rapakivi granite) observed in the >6.0 cm fraction.

total sum of the grains counted. Numerical methods are useful in cases of homogeneous lithologies, where distribution lines for frequencies, based on a negative exponential function (Krumbein 1937, Gillberg 1968) can be used to describe both transport distance and the disintegration rate. Results of analyses of finer material are usually expressed in percentages by weight.

## CASE STUDIES

### The Laitila esker, western Finland

One of the best known pieces of research in this field in Finland was carried out by Hellaakoski (1930), who studied the lithological composition along an esker crossing three main bedrock areas: Jotnian sandstone, rapakivi granite, and metamorphic Proterozoic rocks.

He then compared the material in the esker with that in the surrounding till. The results showed that fragments from a 27 km wide rapakivi granite area appeared in the till close to the bedrock border, but were not entrained in the esker until after a transport distance of about 5 km (Fig. 1). The frequencies were almost twice as high in the esker material as in the till.

From this he drew the conclusion that most of the esker material was derived from the adjacent till. Even though the average transport in the esker was longer than in the till, the glaciofluvial material was more uniform in composition. In this case, when particles have been transported as bedload in glacier meltwater streams, they will be crushed and abraded rapidly, and as a consequence the amount of long-range material remains very low.

### The Hämeenlinna esker, southern Finland

The lithological composition of the Hämeenlinna esker was studied by Virkkala (1958).

A stretch of the esker 38 km long was sampled at 22 points. The esker chain begins from the First Salpausselkä ridge and runs northwestwards through an area of rather variable Proterozoic bedrock. The results of stone counts of three fractions (Ø 4–7 mm, 2–10 cm, and more than 20 cm) were compared with the bedrock composition and the lithological composition of the adjacent till.

The material of the esker indicates slightly longer transportation than for the till. On the average, about 40% of the esker material has its source at a distance more than 5 km. True long-distance transport was observed to be rare. The closest source of Jotnian sandstones, representing the longest transportation distance in the area, is situated at a distance of about 110 km upstream. Sandstones, mainly in the form of small pebbles, were found in nearly all the gravel pits of the area, but these were so few that they did not appear in the stone counts.

The transport distance was found to be dependent on the durability of the rock type, the size of source area and the local topography. Resistant rocks (granites) were transported in noticeable quantities as far as 20 km from their source, whereas the transport distance for less resistant rocks, e.g. metatuffs, seemed to be less than 10 km.

### The glaciofocus method

Analyses of glaciofluvial transport have proved useful for prospecting purposes in an area of Canada covered by thick Quaternary deposits. An esker investigated by Lee (1965) crosses two known gold veins, and the distances from the veins to the esker deposits were determined by studying four mineral species with different hydrological conductivity values and grain sizes. Equivalent transport distances for minerals were then determined from the relationship between grain shape and transport distance under similar hydrodynamic conditions. The results were comparable with those of Hellaakoski (1930).

The maximum frequency of indicator minerals downstream from their source gave a transport distance 'K' of $13 \pm 5$ km for grain sizes 3–8 mm and $5 \pm 3$ km for 8–16 mm (Fig. 2). The 'K' distance was later fitted into an equation and resulted in the glaciofocus method (Lee 1965).

A similar study of a known copper-nickel mineralization was carried out in permafrost areas in Keewatin, Canada, by Shilts (1973). A fixed transport distance was used as a basis, and the results of geochemical analyses from esker material and adjacent tills were compared. The transport distance was explained by the mode of esker sedimentation. Short streams extending back some kilometres from the ice margin had built up esker segments of a few kilometres in length. It was concluded that the expected distance of glaciofluvial transport was about 8 kilometres.

### The Teno valley, Finnish Lapland

Mansikkaniemi (1972) performed stone counts (Ø 2–30 cm) in the valley of the Teno river, Finnish Lapland, to gain a picture of transport distances. He compared the lithological composition of material derived by fluvial transportation with that of glaciofluvially and glacially transported material. An attempt was also made to determine the influence of different transport distances on the roundness of the material. The

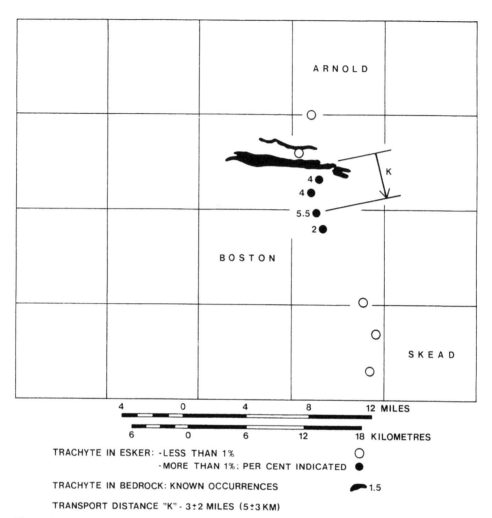

Figure 2. Determination of the K value, after Lee (1965). Transport distance K is the distance from presumed known source in bedrock to position of peak abundance in esker, projected onto a straight line representing average direction of esker.

boundary between the Precambrian and Eocambrian areas which crosses the valley transversally was used in measuring the transport distances.

The results showed that the till in the area was local, 90% of the cobble fraction having moved less than 2 km, whereas about 75% of the glaciofluvial material (esker, deltas) had been transported more than 5 km. Stone counts made at the level of the present river surface showed that this material had moved even further due to recent fluvial activity. About 70% had been transported more than 10 km.

*Pebble lithologies of two eskers in Labrador, Canada*

Bolduc et al. (1987) made lithological analyses of the pebble fraction in order to study

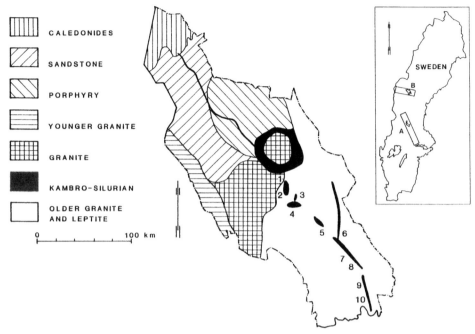

Figure 3. Schematic bedrock map of the county of Dalarna and parts of Västmanland. The black circle denotes the Siljan impact structure. The black lines and figures show the sampling sites along the Badelunda esker. Location A in the index map.

debris entraining, transport and deposition within two esker systems. Stone counts of clast sizes 5 to 20 cm were performed at 28 sites. They were able to conclude that the first appearance and peak percentage (K-value) of certain rock types occur some distance down-esker from where the esker first crossed the bedrock type in question. They also demonstrated that even minor tributary eskers can have a dramatic influence on the composition of a major esker system for a short distance (≤10 km) downstream (see the Badelunda case below).

The pebble lithology of the till adjacent to the esker appeared to reflect changes in the underlying bedrock more rapidly than that of the esker itself. Hence it was concluded that eskers are largely derived from reworked till and not directly from the bedrock.

STUDIES OF THE BADELUNDA AND INDAL VALLEY ESKERS,SWEDEN

*The Badelunda esker*

The 300 km long Badelunda esker (Fig. 3) runs northwestward from the Central Swedish end-moraine zone close to the Baltic (Persson 1975, Möller 1987, Magnusson 1984), through a bedrock area of older granites, veined gneisses and acid volcanics (SGU map No. Ah 7, 1984). In the NW part of the county of Dalarna, the provenance area of the Weichselian glaciers, the esker contains porphyries and sandstones, in accordance with the bedrock lithology (G. Lundqvist 1951, G. Lundqvist and Hjelmqvist 1941, 1946,

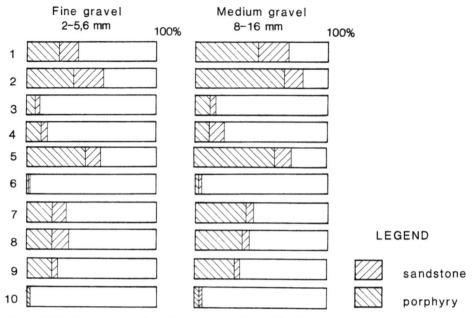

Figure 4. Lithology variation along the Badelunda esker and its tributaries.

Hjelmqvist and G. Lundqvist 1953, Kulling and Hjelmqvist 1948). These rock types are easily detected in glacial deposits settled by ice and water.

The results of stone counts are shown in Figure 4. It can be seen that the tributaries (nos. 3 and 6) contain a lower frequency of long-distance transported porphyries and sandstones in their gravel fractions than does the main esker stream. Comparison between the fine and medium-size gravels of the indicator clasts does not show any increase in the proportion of the fine gravel fraction with increased length of transport. Instead the fine gravels seem to decrease, and although not shown here, both fractions disappear more or less simultaneously.

The percentage of porphyries decreases downstream from the confluence of each tributary due to dilution caused by incoming material of other lithologies. After a glaciofluvial transport of about 10 km the frequency of porphyries rises again, and it remains at about 70% as far as 90 km downstream from the source. Here the percentage falls rapidly. This indicates transport distances of at least 100 km, and occasionally up to 150 km. The sandstones have lower percentages along the esker, but they reach higher amounts in the sand fraction.

G. Lundqvist and Hjelmqvist (1941) found by stone counting that the Svärdsjö esker, earlier believed to be the main path, was in fact a tributary to the northwest trending Badelunda esker. The two tributaries run through the same bedrock lithology but they have different drainage areas.

The lithologies found in the eskers at two juxtaposed excavations some kilometres upstream from the confluence area, were quite different. The Badelunda esker contains about 50% long-distance material compared with less than 10% in the Svärdsjö esker, which corresponds to the amount in the surrounding till.

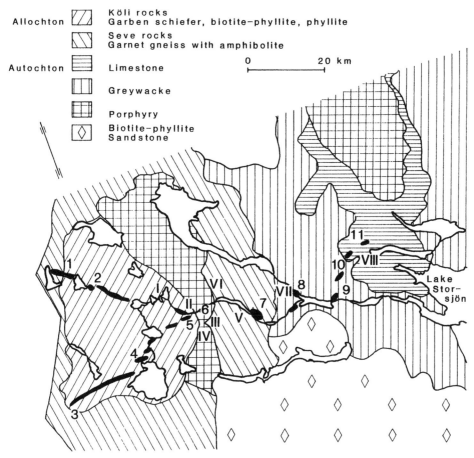

Allochton — Köli rocks
Garben schiefer, biotite–phyllite, phyllite

Seve rocks
Garnet gneiss with amphibolite

Autochton — Limestone

Greywacke

Porphyry

Biotite–phyllite
Sandstone

0          20 km

Lake Stor-sjön

Figure 5. Combined schematic bedrock map covering the Indal valley esker stretch. Numbers 1–11 denote sampling sites in the esker and Roman numerals I–VIII sampling sites in the till. Location B in Figure 3.

A further conclusion was that entrained till from nearby would result in a similar lithological composition in the two branches. This difference indicated an extremely long transportation in glaciofluvial mode.

*The Indal valley esker*

The Indal valley esker (Fig. 5), beginning with two branches in the Caledonian mountains close to the Norwegian border, runs eastward more or less continuously in a ridge form. After confluence, the scattered deposits follow the Indal river valley, clustering against the valley sides. Closer to Lake Storsjön, the esker changes to a northerly direction, withdrawing into a hilly landscape. The deposits contain glaciofluvially sorted material in uphill positions and unsorted morainic material on the northern sides of the hills (J. Lundqvist 1969).

According to the bedrock map the esker crosses four bedrock lithologies (SGU map

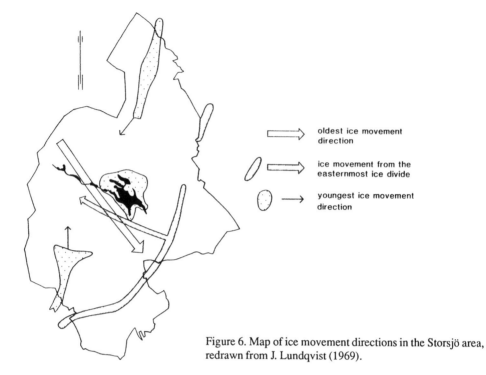

Figure 6. Map of ice movement directions in the Storsjö area, redrawn from J. Lundqvist (1969).

No. Ca 53, 1984). The Köli nappe in the westernmost part of the area contains Garben schiefers and phyllites. The underlying Seve nappe has gneiss and amphibolite facies, and between these nappes is a 'window' of porphyries. Further east is a vast area of limestones and greywackes. All of these bedrock lithologies contains calcareous grey-wackes.

The esker runs through an area where the ice divide shifted. After an older ice movement emanating from the mountains in the west, the ice reached a maximum position east of Lake Storsjön (Frödin 1914, Ljungner 1949) (Fig. 6). During final stages of deglaciation the ice disintegrated and split up into several smaller lobes in the valleys. The ice in the Indal valley withdrew eastwards, and lakes were dammed up by the mountains. The esker developed subglacially in these lakes (J. Lundqvist 1969). Finally the ice became inactive in the Storsjö basin, changing the drainage pattern through the stagnant ice towards the east.

Stone counts have been performed on 2–5.6 mm and 8–20 mm fractions. The indicator lithologies are shown in Figure 5. There were some complicating factors. Many of the shales are calcareous and therefore difficult to differentiate from limestones, and it is thus possible that the limestone content may be high due to a local bedrock effect. This can be seen in Figure 7 (sample 7) in a fluctuating limestone peak which does not increase in any direction.

The till contains a higher percentage of the fragile Garben schiefer than the esker material, within as well as outside the source area. Within the provenance area the close relationship between the bedrock and the till is obvious, as demonstrated earlier by

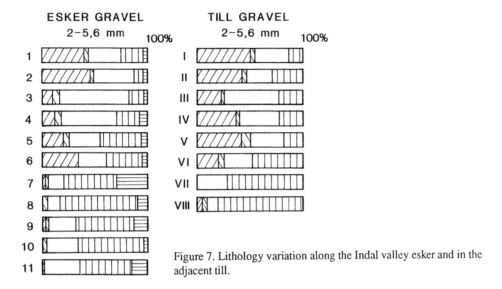

Figure 7. Lithology variation along the Indal valley esker and in the adjacent till.

Lindén (1975) and Perttunen (1977) in other areas. The higher frequencies in the till outside the source area are probably due to ice crushing, concentrating the easily disintegrated rock fragments in the gravel-size fraction. In the esker this fragile material is quickly abraded into finer fractions and diluted by fresh material.

The Garben schiefers and phyllites of the Köli nappe are seen 40 km east of the bedrock border. Since their only source is in the far west, the transport distance must be in accordance with this length. The frequency curve of the greywackes is more difficult to interpret. If they were deposited from the ice as it withdrew eastwards, the frequency curve is in agreement with the results of Hellaakoski (1930) for the source area. Further west the high amounts of greywackes in the Köli nappe area are more likely to reflect local abundance than a decreasing frequency from the east.

In spite of two periods of glacial transport before transportation in a glaciofluvial system, the easily disintegrating Garben schiefers have resisted total disintegration. Their low frequency in the esker material may be due to the grains being exposed to water and frost action – processes which quickly split and comminute the fragments. The uncertain curves for the limestones and greywackes are not a reliable indicator of transport in this area. The results so far indicate pronounced transport from the ice maximum in the west and a late, less erosive glacial phase, and in consequence, weak transport from the ice maximum in the east (Lilliesköld 1986).

CONCLUSIONS

The method of tracing bedrock fragments along glaciofluvial systems has evident applications. In poorly known areas, stone counts can be used to complement other methods used for indicating ice flow directions. Gravel counts are also a fast and easy way of gaining general information on the bedrock of an area. In drift-covered areas with sparsely exposed bedrock, esker material has proved useful in mineral exploration. A kimberlite belt was found in Canada using the glaciofocus method developed by Lee (1965, 1968), for example. The 'K' value, i.e. the maximum frequency of indicator minerals downstream from their source, gave a transport distance of $5 \pm 3$ km for the coarser gravel and $13 \pm 5$ km for the finer gravel fractions, respectively.

Glaciofluvial transport distances vary according to the energy level of the glacial meltwater system, the grain size studied and the resistance of the indicator rocks. The grain size and shape of the particles does not only express the transport distance but also the different comminution rates specific for the rock types. The soft fragments of the Indal valley esker, for example, seemed to be transported in glaciofluvial mode for a distance of 40 km, whereas in the Badelunda esker the proportion of hard porphyries remains at about 70% as far as 90 km downstream.

On the basis of some Finnish esker studies the expected average distance from the provenance area is about five to eight kilometres. Particles within esker systems transported as bedload in glacial meltwater streams have been crushed and abraded rapidly, and in consequence the true amount of long-distance material remains low. Even though such pebbles can be found in nearly all gravel pits, their frequency is usually insignificant, of the order of 0.1% to 0.01%.

REFERENCES

Banerjee, I. and B.C. McDonald, 1975. Nature of esker sedimentation. In: Jopling and B.C. McDonald (eds.), Glaciofluvial and glaciolacustrine sedimentation. Society of Paleontologists and Mineralogists. Special publication 23: 132–154.

Bolduc, A.M., R.A. Klassen and E.B. Evenson, 1987. Cobble lithologies in eskers of central Labrador. Current Research, Part A, Geological Survey of Canada, Paper 87–1A: 43–51.

Boulton, G.S., 1979. Processes of glacier erosion on different substrata. Journal of Glaciology 23(89): 15–38.

Boulton, G., D.L. Dent and E.M. Morris, 1974. Subglacial shearing and crushing, and the role of water pressures in tills from south-east Iceland. Geografiska Annaler 56 A: 121–134.

Drewry, D., 1986. Glacial Geological Processes. Edward Arnold, 276 pp.

Evenson, E.B. and J.M. Clinch, 1987. Debris transport mechanism at active Alpine glacier margins: Alaskan case studies. In: R. Kujansuu and M. Saarnisto (eds.), INQUA Till Symposium. Geological Survey of Finland, Special Paper 3: 111–136.

Flint, R.F., 1971. Glacial and Quaternary Geology. John Wiley and Sons, New York, 892 pp.

Frödin, G., 1914. Hufvuddragen af isavsmältningen inom nordvästra Jämtland. Geologiska Föreningen i Stockholm Förhandlingar 36: 131–156.

Gillberg, G., 1968. Lithological distribution and homogeneity of glaciofluvial material. Geologiska Föreningen i Stockholm Förhandlingar 90: 189–204.

Gumaelius, O., 1885. Stenräkning i Uppsala- och Enköpingsåsarne. Geologiska Föreningen i Stockholm Förhandlingar 7: 777–788.

Gummaelius, O., 1871. Några ord till upplysning om bladet Engelsberg. Sveriges Geologiska Undersökning, Ser. Aa 42, 50 pp.

Gurnell, A.M. and M.J. Clark, 1987. Glacio-fluvial sediment transfer: An alpine perspective. John Wiley & Sons, New York.

Hallet, B., 1979. A theoretical model of glacial abrasion. Journal of Glaciology, 23(89): 39–50.

Hellaakoski, A., 1930. On the transportation of materials in the esker of Laitila. Fennia 52, 42 pp.

Hjelmqvist, S., 1966. Description to map of the pre-Quaternary rocks of the Kopparberg county, Central Sweden. Sveriges Geologiska Undersökning Ca 40, 217 pp.

Hjelmqvist, S. and G. Lundqvist, 1953. Beskrivning till kartbladet Säter. Sveriges Geologiska Undersökning Aa 194, 97 pp.

Krumbein, W.C., 1937. Sediments and exponential curves. Journal of Geology 45: 577–601.

Kulling, O. and S. Hjelmqvist, 1948. Beskrivning till kartbladet Falun. Sveriges Geologiska Undersökning, Ser. Aa 189, 184 pp.

Lee, H.A., 1965. Investigations of eskers for mineral exploration. Geological Survey of Canada, Paper 65–14, 20 pp.

Lee, H.A., 1968. An Ontario kimberlite occurrence discovered by application of the Glaciofocus method to a study of the Munro esker. Geological Survey of Canada, Paper 68–7, 3 pp.

Lilliesköld, M., 1986. Köliskiffrarnas transport. Abstract till 17e Nordiska Geologmötet, Helsingfors.

Lilliesköld, M., 1990: A method using Image Analysis for Grain-shape Measurements. Striae 30.

Lindén, A., 1975. Till petrographical studies in an Archaean bedrock area in southern central Sweden. Striae 7, 57 pp.

Ljungner, E., 1949. East-west balance of the Quaternary ice caps i Patagonia and Scandinavia. Bulletin of the Geological Institute, University of Uppsala 33: 11–97.

Lundqvist, G., 1951. Beskrivning till jordartskarta över Kopparbergs län. Sveriges Geologiska Undersökning Ca 21, 213 pp.

Lundqvist, G. and S. Hjelmqvist, 1941. Beskrivning till kartbladet Hedemora. Sveriges Geologiska Undersökning, Der. Aa 184: 74–83.

Lundqvist, G. and S. Hjelmqvist, 1946. Beskrivning till kartbladet Avesta. Sveriges Geologiska Undersökning, Ser. Aa 188, 127 pp.

Lundqvist, J., 1952. Bergarterna i Dalamoränernas block- och grusmaterial. Sveriges Geologiska Undersökning C 525, 48 pp.

Lundqvist, J., 1969. Beskrivning till jordartskarta över Jämtlands län. Sveriges Geologiska Undersökning Ca 45, 418 pp.

Magnusson, E., 1984. Description to the Quaternary map Västerås SE. Sveriges Geologiska Undersökning, Ser Ae 64, 76 pp.

Mansikkaniemi, H., 1972. Flood deposits, transport distances and roundness of loose material in the Tana river valley, Lapland. Kevo Subarctic Research Station, Report 9: 15–23.

Möller, H., 1987. Description to the Quaternary map Strängnäs SW. Sveriges Geologiska Undersökning, Ser. Ae 82, 60 pp.

Okko, V., 1945. Untersuchungen über den Mikkeli-Os. Fennia 69: 1-55.

Persson, Ch., 1975. Description to the Quaternary map Nyköping NE. Sveriges Geologiska Undersökning, Ser. Ae 21, 83 pp.

Perttunen, M., 1977. The lithologic relation between till and bedrock in the region of Hämeenlinna, southern Finland. Geological Survey of Finland, Bulletin 291, 68 pp.

von Post, H., 1855. Om sandåsen vid Köping i Westmanland. Kongl. Svenska Vetenskaps-Akademiens Handlingar 1854, Stockholm 1856, 345 pp.

Powers, M.C., 1953. A new roundness scale for sedimentary particles. Journal of Sedimentary Petrology 23: 117–119.

Rice, R.J., 1982. The hydraulic geometry of the lower portion of the Sunwapta River valley train, Jasper National Park, Research in glacial-glaciofluvial and glaciolacustrine systems. Proceedings of 6th Guelph Symposium on Geomorphology 1980. Geo Books: 151–173.

Röthlisberger, H. and A. Iken, 1981. Plucking as an effect of water pressure variations at the glacier bed. Annals of Glaciology 2: 57–62.

Shilts, W., 1973. Drift prospecting; geochemistry of eskers and till in permanently frozen terrain: District of Keewatin; Northwest Territories. Geological Survey of Canada, Paper 72–45, 34 pp.

Slatt, R.M. and N. Eyles, 1981. Petrology of glacial sand: implications for the origin and mechanical durability of lithic fragments. Sedimentology 28: 171–183.

Stone, G., 1899. The glacial gravels of Maine. U.S. Geological Survey, Monograph 34, 499 pp.

Trefethen, J.M. and H.B. Trefethen, 1944. Lithology of the Kennebec valley esker. American Journal of Science 242: 521–527.

Virkkala, K., 1958. Stone counts in the esker of Hämeenlinna, Southern Finland. Bulletin de la Commission géologique de Finlande 180: 87–103.

Zingg, T., 1935. Beitrag zur Schotteranalyse. Mineralogische und Petrographische Mitteilungen 15: 39–140.

Maps
Geological Survey of Sweden, Bedrock map 1:250 000, Ah 7, 1984.
Geological Survey of Sweden, Bedrock map 1:200 000, Ca 53, 1984.

# CHAPTER 10

# Heavy minerals in glacial material

VESA PEURANIEMI

*Department of Geology, University of Oulu, Finland*

## INTRODUCTION

The heavy minerals are generally regarded as comprising those minerals which have a high specific gravity, usually higher than that of the more common rock-forming minerals such as quartz and the feldpars. According to Bates and Jackson (1980), the specific gravity of the heavy minerals is over $2.85 - 2.9$. Heavy mineral investigations quite often concentrate on minerals which have a specific gravity of well over 3.0, as most ore minerals have, but it must be remembered that not all ore minerals are heavy, e.g. beryl $(Be_3Al_2 Si_6O_{18})$ with a specific gravity of $2.6 - 2.9$, or petalite $(LiAlSi_4O_{10})$, the specific gravity of which is $2.3 - 2.5$. Kuzin and Egorov (1976) divide natural minerals into four groups according to their specific gravities: 1. light minerals, s.g. under 2.5; 2. intermediate minerals, s.g. $2.5 - 3.3$; 3. heavy minerals, s.g. $3.4 - 6$; and 4. very heavy minerals, s.g. over 6. This classification groups minerals such as tourmaline, spodumene and apatite, which are generally regarded heavy, together with quartz and the feldspars.

The weathering-resistant oxide and silicate minerals are usually used in heavy-mineral studies, while the more labile sulphide minerals have been studied in stream sediments but seldom in till deposits. Sulphide minerals may have survived in till in certain environments, however, and so they may also be applicable in heavy-mineral till studies. Heavy minerals are very useful for tracing till provenance, examining till stratigraphy and prospecting for ores.

## EARLIER STUDIES

Numerous heavy-mineral studies using samples from stream sediments have been performed in many parts of the world (e.g. Theobald and Thompson, 1959; Ossenkopf et al., 1979; Tencik and Sponar, 1979; Callahan, 1981; Watters, 1983; Zantop and Nespereira, 1979; Dunlop and Meyer, 1978; Day and Fletcher, 1986; Saxby and Fletcher, 1986). Mertie (1979) used the heavy-mineral fraction of saprolite (deeply weathered bedrock) for geological exploration.

Lehmuspelto (1976) studied the dispersion of cassiterite in till using large samples (one sample of 200 kg), and Kokkola and Pehkonen (1976) also used large samples (one sample 150 – 170 1) to study the transport of gold in till. Nikkarinen and Björklund (1976a-b) tested the use of small samples taken with a percussion drill (one sample 200 g) in their heavy-mineral exploration for spodumene and scheelite. Lindmark (1977) used

Table 1. The most common heavy minerals and their properties.

| Mineral | Formula | Specific gravity | Hardness (Mohs scale) | Magnetic suscepti-bility |
|---|---|---|---|---|
| sulphides | | | | |
| chalcopyrite | $CuFe_2$ | 4.1-4.3 | 3-4 | nm |
| sphalerite | $ZnS$ | 3.9-4.1 | 3-4 | nm |
| galena | $PbS$ | 7.4-7.6 | 2-3 | nm |
| pentlandite | $(Fe,Ni)_9S_8$ | 4.5-5.0 | 3-4 | nm |
| molybdenite | $MoS_2$ | 4.6-5.0 | 1-1.5 | nm |
| arsenopyrite | $FeAsS$ | 5.9-6.2 | 5.5-6 | nm |
| pyrite | $FeS_2$ | 4.9-5.2 | 6-6.5 | nm |
| pyrrhotite | $FeS$ | 4.6-4.7 | 3.5-4.5 | wm |
| oxides | | | | |
| cassiterite | $SnO_2$ | 6.8-7.0 | 6-7 | nm |
| chromite | $FeCr_2O_4$ | 4.0-4.8 | 5.5-7.5 | wm |
| gahnite | $ZnAl_2O_4$ | 4.0-4.6 | 7.5-8 | nm |
| magnetite | $Fe_3O_4$ | 4.9-5.2 | 5.5-6.5 | sm |
| hematite | $Fe_2O_3$ | 4.9-5.3 | 5.5-6.5 | wm |
| ilmenite | $FeTiO_3$ | 4.5-5.0 | 5-6 | wm |
| rutile | $TiO_2$ | 4.2-4.3 | 6.0-6.5 | nm |
| goethite | $FeO(OH)$ | 4.0-4.4 | 4.5-5.5 | wm |
| uraninite | $UO_2$ | 9.0-9.7 | 5.0-5.5 | nm |
| columbitetantalite | $(Fe,Mn,Mg)(Nb,Ta,Sn)_2O_6$ | 5.2-8.2 | 4.2-7.0 | |
| phosphates | | | | |
| apatite | $Ca_5(PO_4)_3(F,OH,Cl)$ | 3.2 | 5 | wm |
| monazite | $(Ce,La,Y,Th)PO_4$ | 4.9-5.5 | 5.0-5.5 | wm |
| wolframates | | | | |
| scheelite | $CaWO_4$ | 5.8-6.2 | 4.5-5 | nm |
| wolframite | $(Fe,Mn)WO_4$ | 7.0-7.5 | 5.0-5.5 | wm |
| sulphates | | | | |
| jarosite | $KFe_3(SO_4)_2(OH)_6$ | 3.1-3.2 | 2.5-3.5 | nm |
| anglesite | $PbSO_4$ | 6.1-6.4 | 2.5-3.0 | nm |
| baryte | $BaSO_4$ | 4.3-4.5 | 3.0-3.5 | nm |
| carbonates | | | | |
| cerussite | $PbCO_3$ | 6.4-6.6 | 3.0-3.5 | nm |
| smithsonite | $ZnCO_3$ | 4.1-4.5 | 5 | nm |
| halides | | | | |
| fluorite | $CaF_2$ | 3.0-3.2 | 4 | nm |
| elements | | | | |
| gold | $Au$ | 15.6-19.3 | 2.5-3.0 | nm |
| silicates | | | | |
| amphiboles | $A_2B_5(Si,Al)_8O_{22}(OH)_2$ $A = Mg,Fe^{2+},Ca,Na$ $B = Mg,Fe^{2+},Al,Fe^{3+}$ | 2.8-3.4 | 6 | wm |
| pyroxenes | $ABSi_2O_6$ $A = Mg,Fe^{2+},Ca,Na$ $B = Mg,Fe^{2+},Al$ | 3.1-3.6 | 5-7 | wm |
| olivine | $(Mg,Fe)_2SiO_4$ | 3.2-3.5 | 6.5-7.0 | wm |
| biotite | $K(Mg,Fe,Mn)_3(Al,Fe)Si_3O_{10}(OH,F)_2$ | 2.7-3.1 | 2.5-3.0 | wm |
| chlorite | $(Mg,Fe)_3(Si,Al)_4(OH) \cdot (Mg,Fe)_3(OH)_6$ | 2.6-3.4 | 2.0-2.5 | wm |

Table 1. The most common heavy minerals and their properties (cont.).

| Mineral | Formula | Specific gravity | Hardness (Mohs scale) | Magnetic suscepti- bility |
|---|---|---|---|---|
| silicates | | | | |
| garnets | $A_3B_2(SiO_4)_3$ $A = Ca,Mg,Fe^{2+},Mn$ $B = Al,Fe^{3+},Cr,V$ | 3.5-4.2 | 6.0-7.5 | wm |
| epidote | $Ca_2Al_2FeOSiO_4Si_2O_7(OH)$ | 3.3-3.5 | 6.5 | wm |
| zircon | $ZrSiO_4$ | 4.7 | 7-8 | nm |
| topaz | $Al_2SiO_4(F,OH)_2$ | 3.4-3.6 | 8 | nm |
| tourmaline | $Na(Fe,Mn)_3Al_6B_3Si_6O_{27}(OH,F)_4$ | 3.0-3.2 | 7.0-7.5 | wm |
| titanite | $CaTiSiO_4$ | 3.4-3.6 | 5.0-5.5 | wm |
| spodume | $LiAlSi_2O_6$ | 3.0-3.2 | 6.5-7.0 | nm |

nm= non-magnetic
wm= weakly magnetic
sm= strongly magnetic

various sample sizes (200 g – 8000 g) when studying the transport of scheelite in till, while Brundin and Bergström (1977) performed a thorough investigation on the concentration and analysis of heavy minerals in till using a sample size of 10 1. They have also studied the dispersion of scheelite. Toverud in his many papers (1979, 1982, 1984) has described the transport of cassiterite and scheelite in till using a sample size of 5 1. Thompson (1979) applied the heavy-mineral geochemistry of till to sulphide prospecting, and Peuraniemi (1981, 1984, 1987) studied the dispersion of cassiterite using a sample size of 10 1/one sample. Closs and Sado (1982) used the heavy-mineral fractions of small till samples in their carbonatite prospecting, while Saarnisto and Tamminen (1985, 1987) studied the occurrence of gold in the heavy-mineral concentrates of till and glaciofluvial material using large samples (50 1).

Heavy minerals have been used in till provenance and till stratigraphy studies by Dreimanis and Reavely (1953), DiLabio and Shilts (1979), Evenson et al. (1979) and Peuraniemi (1982). Heavy minerals occurring in sorted glaciofluvial sediments such as eskers have been studied relatively little (Lee, 1963, 1965, 1968).

PHYSICAL AND CHEMICAL PROPERTIES OF THE HEAVY MINERALS

The minerals most commonly studied in heavy-mineral geochemistry and their formulae, specific gravity, hardness and magnetic susceptibility are presented in Table 1. We can see, for example, that the hardness of most sulphide minerals is clearly lower than that of oxide and silicate minerals, so that they are weaker in their resistance to mechanical weathering and their grain size diminishes rapidly in glacial milling. Although some minerals such as scheelite are relatively hard, they can be brittle, and therefore also rather weak in terms of mechanical weathering.

All the minerals that have been crystallized deep in earth's crust are unstable under

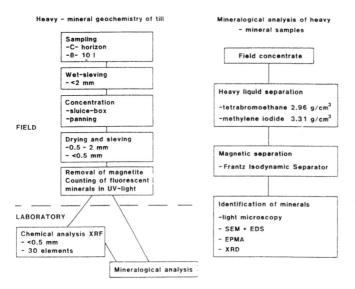

**Heavy - mineral geochemistry of till**

Sampling
-C- horizon
-8- 10 l

Wet-sieving
- <2 mm

Concentration
-sluice-box
-panning

FIELD

Drying and sieving
-0.5 - 2 mm
- <0.5 mm

Removal of magnetite
Counting of fluorescent
minerals in UV-light

LABORATORY

Chemical analysis XRF
- <0.5 mm
- 30 elements

Mineralogical analysis

**Mineralogical analysis of heavy - mineral samples**

Field concentrate

Heavy liquid separation
-tetrabromoethane 2.96 g/cm³
-methylene iodide 3.31 g/cm³

Magnetic separation
- Frantz Isodynamic Separator

Identification of minerals
-light microscopy
- SEM + EDS
- EPMA
- XRD

Figure 1. Flow sheets of the heavy mineral geochemistry and the mineralogical techniques used in heavy mineral analysis.

conditions prevailing on the surface, and all minerals weather chemically, some rapidly and some so slowly that they give the impression of being unweatherable. Oxides are most slowly weathering of the heavy minerals, and wolframates also weather very slowly. Thus these mineral groups are generally the most useful for heavy-mineral studies.

On the other hand, sulphides and carbonates weather readily in acid and oxidizing conditions. Among the silicate minerals the dark Fe-Mg minerals are heavy, and they are also the most unstable silicates in terms of chemical weathering. Zircon is usually a resistant mineral, but weathers under the influence of bicarbonate-rich water (Carroll, 1953).

SAMPLING

The till samples for heavy-mineral surveys are most often taken with a shovel from the surficial part of till. When samples are needed from deeper parts a tractor excavator is used. With these two sampling methods it is possible to obtain large samples. Various drilling methods, e.g. percussion, pneumatic and auger drilling have also been used, but they all have the common disadvantage that the sample size is small. Thompson (1979) used a dual-tube reverse circulation rotary drill to get continuous till samples from beneath thick glaciolacustrine deposits. In the case studies reported here a shovel and a tractor excavator were used.

CONCENTRATION

Since most heavy minerals are already accessories in the bedrock, they occur sparsely in glacial sediments. It is thus necessary to concentrate the samples in order to facilitate the study of heavy minerals.

The oldest concentration method available is panning (Mertie, 1954), which has been used especially in the search for gold. The advantages of panning are its simplicity and cheapness, but one of the disadvantages is its slowness for handling large samples. Only a pan and water are needed, just as water is the only medium needed in a sluice box with riffles on its floor. The sluice box has also been commonly used in the search for gold. In old times long wooden boxes were used but nowadays they are shorter and of metal, and are combined with a combustion motor-driven suction pump (Brundin and Bergström, 1977). Sluicing is a fast concentration method, but is generally insufficient on its own. The concentrate from the sluice box may be further concentrated by other methods (Peuraniemi, 1987). Heavy liquids such as bromoform, tetrabromoethane, methylene iodide and Clerici solution are used in laboratories to achieve a very exact separation of heavy minerals according to their specific gravities (Reeves and Brooks, 1978). The use of these heavy liquids is slow and expensive, however, and thus they are uneconomic for to the handling of large sample quantities.

The concentrating table is one possible method, both in the laboratory and in a field laboratory (e.g. Lindmark, 1977). Magnetic separation in its simplest form consists of the removal of magnetite by means of a hand magnet. Another form of magnetic separation is the use of an Eclipse 'Quick-Release' hand magnet (Jones and Fleming, 1965). A better separation based on magnetic susceptibility can be obtained with a Frantz Isodynamic Separator. Magnetic separation is used mostly as a supplement to other concentration methods. The examples contained in this article employ panning, sluicing, heavy liquids, a hand magnet and the Frantz Isodynamic Separator (Fig. 1).

ANALYSIS AND IDENTIFICATION

Heavy-mineral concentrates are usually first analyzed chemically, by X-ray fluorescence spectrometry (XRF), atomic absorption spectrometry (AAS), optical emission spectrometry (OES) or instrumental neutron activation analysis (INAA) (Fletcher, 1981). Since chemically quite stable minerals are concerned, they must be totally dissolved for AAS analysis. In general 20 – 40 major, minor and trace elements are analyzed. XRF is a highly recommendable method for heavy-mineral samples, because it is fast and employs grinding as its only form of sample preparation.

Many methods can be used to identify heavy minerals. Some, e.g. gold, are identifiable by the naked eye, while a more advanced method employs a stereomicroscope to examine the minerals as such without any preparation. Quantitative mineral counts can be performed quite quickly with a stereomicroscope, but only an experienced geologist can reliably identify heavy minerals only on the basis of colour and morphology. Some help is usually needed, e.g. through the use of a staining technique, for example the 'tinning' of cassiterite grains (von Philipsborn, 1967). Some minerals as scheelite and zircon are fluorescent under short-wave ultraviolet light, while spodumene grains show a brown colour and a 'microcrack' appearance when heated to 1000°C and are thus easier to identify in that state.

Heavy-mineral concentrates can be mounted in Epofix resin for the production of thin sections, polished thin sections or polished sections, which can then be studied qualitatively and quantitatively with a polarizing microscope in transmitted or reflected light.

All the minerals in a concentrate can in principle be identified by X-ray diffraction. If a

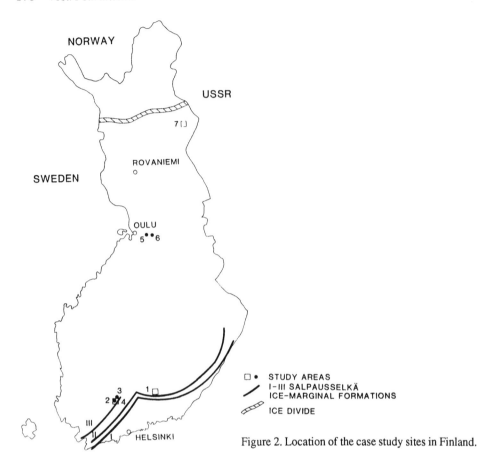

Figure 2. Location of the case study sites in Finland.

concentrate includes numerous different minerals, it is possible that the peaks in the diffractogram may overlap, making identification difficult or even impossible.

The most modern means for identifying heavy minerals are electron optical devices. Minerals, their alteration products and surface textures can be identified quite easily and quickly using a scanning electron microscope equipped with an energy-dispersive X-ray spectrometer (SEM+EDS). One mineral grain can be analyzed semiquantitatively by EDS. The exact chemical composition of minerals may be determined using an electron probe microanalyzer (EPMA). Either a polished section or a polished thin section from a heavy-mineral concentrate is needed for EPMA, while separate mineral grains as such can be studied by SEM. Thompson and Hale (1984) used laser ablation combined with inductively coupled plasma emission spectrography to identify heavy minerals. XRF, light microscopes, XRD, SEM+EDS and EPMA have been used in the present examples (Fig. 1).

CASE STUDIES

The locations of the case study sites in Finland are shown in Figure 2.

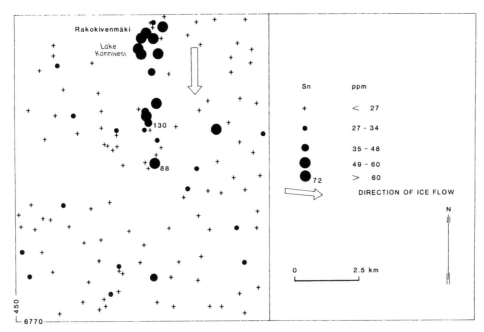

Figure 3. Tin content in heavy mineral samples from till at Rakokivenmäki.

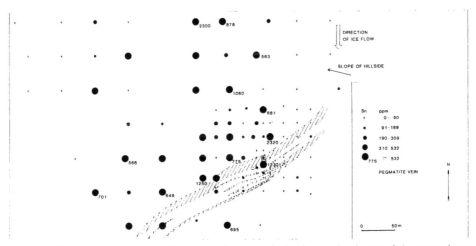

Figure 4. Tin content in heavy mineral samples from till in the proximal part of the anomaly at Rakokivenmäki.

## 1. *Rakokivenmäki Sn*

The site is located in an area of Proterozoic granitic rocks 6 km west of the Viipuri rapakivi massif in southern Finland. One till bed, lodgement till in its facies, partly covers the bedrock. The landscape type is a cover moraine. The last deglacial ice flow occurred

Table 2. Mineralogical composition of three heavy-mineral samples from Rakoki-venmäki.

| 300 ppm Sn | 160 ppm Sn | 130 ppm Sn |
|---|---|---|
| hematite | cassiterite | almandite |
| amphibole | zircon | amphibole |
| almandite | rutile | hematite |
| cassiterite | sillimanite | ilmenite |
| quartz | scheelite | chlorite |
| | fluorapatite | tourmaline |
| | | epidote |
| | | quartz |
| | | plagioclase |
| | | mica |

Table 3. Chemical composition of seven cassiterite grains in three heavy-mineral samples from Rakokivenmäki (in wt%).

| | 300 ppm Sn | 160 ppm Sn | | | 130 ppm Sn | | |
|---|---|---|---|---|---|---|---|
| | 1 | 2 | 3 | 4 | 5 | 6 | 7 |
| MgO | 0.1 | 0.4 | 0.4 | 0.1 | 0.1 | 0.1 | 0.1 |
| $Al_2O_3$ | 0.1 | 0.1 | 0.2 | 0.1 | 0.2 | 0.1 | 0.1 |
| $TiO_2$ | 0.4 | 0.6 | 0.2 | 0.1 | 0.1 | 0.1 | 0.4 |
| $Fe_2O_3$ | 0.8 | 0.5 | 0.6 | 0.4 | 0.4 | 0.3 | 1.1 |
| MnO | 0.1 | 0.1 | 0.2 | 0.1 | 0.1 | 0.1 | 0.2 |
| $Nb_2O_5$ | 0.9 | 0.7 | 0.8 | 0.3 | 0.2 | 0.2 | 2.0 |
| $Ta_2O_5$ | 2.2 | 0.4 | 1.2 | 0.8 | 0.9 | 0.8 | 1.7 |
| $SnO_2$ | 93.2 | 97.5 | 97.3 | 98.3 | 96.8 | 96.8 | 94.1 |
| Sum | 97.8 | 100.3 | 100.9 | 100.2 | 98.8 | 98.5 | 99.7 |

from north to south, towards the Second Salpausselkä ice-marginal complex. The tin content of the heavy-mineral concentrates of till is presented in Figure 3. A quite coherent rectilinear anomaly occurs in the direction of ice movement. The anomaly is 5 km in length. The Second Salpausselkä ice-marginal formation lies 2.5 km south of the distal end of the tin anomaly. The sampling grid has been made denser in the proximal part of the anomaly (Fig. 4), which is situated on a hillside which slopes quite steeply to the west and west-northwest. Mass movements may have occurred on the slope after the water-logged till had been deposited, and it is probably this that caused the westward shift in the anomaly. The shore of Lake Konnivesi forms the western limit of the anomaly.

The mineralogy of three heavy mineral samples from the proximal, middle and distal parts of the anomaly and the chemical composition of their cassiterite are presented in Tables 2-3. Both the mineralogy and the high Nb and Ta-concentrations in the cassiterite suggest that the source rock of the tin anomaly should be a complex pegmatite (Dudykina, 1959). Later several complex pegmatite veins with an Sn content of 400 – 500 ppm were also found in the proximal part of the anomaly (Fig. 4).

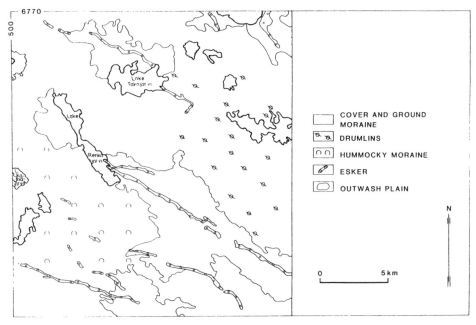

Figure 5. Glacial geology of the area west of Hämeenlinna, southern Finland.

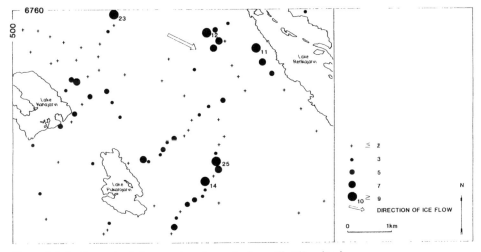

Figure 6. Scheelite grains in heavy mineral samples from till at Kanajärvi.

## 2. *Kanajärvi W, Sn, Ta*

Kanajärvi is situated 27 km west of the town of Hämeenlinna in southern Finland. The bedrock is composed of Proterozoic granitoids and mica schists. The area is a vast hummocky moraine field near outwash plains belonging to the Third Salpausselkä ice-marginal formation (Fig. 5). The last direction of ice movement was from the

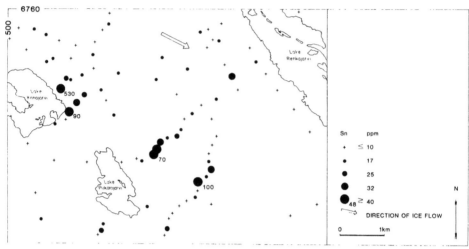

Figure 7. Tin content in heavy mineral samples from till at Kanajärvi.

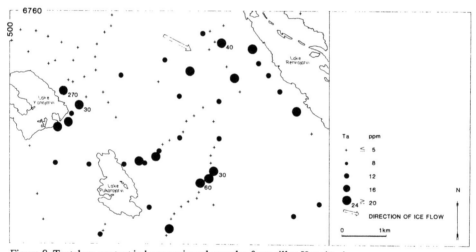

Figure 8. Tantalum content in heavy mineral samples from till at Kanajärvi.

west-northwest. The moraine hummocks are composed mainly of a grey, sandy till. The numbers of scheelite grains in the heavy-mineral till samples are shown in Figure 6. Two separate anomaly fans are to be seen, both oriented in the direction of ice movement. The bedrock source of the northern anomaly (length 4 km, breadth in the distal part 600 m) is not known. Scheelite dissemination in the skarn-banded volcanics and chert horizon has been found in the proximal part of the southern anomaly (length 4 km, breadth 1.2 km in the proximal part and 2 km in the distal part). This source rock is not shown in Figure 6.

Tin and tantalum anomalies occur in the same area as the southern scheelite anomaly (Figs. 7-8), but are ribbon-like rather than fan-like. A test pit (Fig. 9) was dug at the proximal end of the tin anomaly (530 ppm Sn) with an excavator. Topmost in the profile is

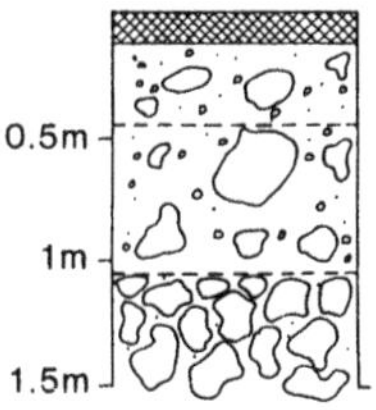

| | HEAVY-MINERAL CONCENTRATE <0.5mm ppm | | | FINE FRACTION OF TILL | |
|---|---|---|---|---|---|
| | | | | <0.06 mm ppm | 0.06 -0.25 mm ppm |
| | Sn | Ta | Nb | Sn | Sn |
| HUMUS | | | | | |
| B-HORIZON | 130 | 0 | 10 | | |
| HOMOGENEOUS, GREY SANDY TILL | 530 480 | 0 10 | 20 30 | 6 | 4 |
| MIXTURE OF TILL AND WEATHERED BEDROCK | 1310 | 270 | 880 | 10 | 15 |

Figure 9. A test pit in the till at Kanajärvi.

a rather thick humus ($A_0$) horizon (10 cm). No eluviation horizon ($A_2$) can be discerned. The thickness of the illuviation horizon (B) is 35 cm. The parent material is a homogeneous grey sandy till. A mixture of till and weathered bedrock occurs at a depth of 1.05 m. The weathered bedrock is composed of coarse-grained reddish pegmatite. Because of the compactness of the material, excavation work had to be discontinued at a depth of 1.5 m. Sn, Ta and Nb increase from the top to the bottom in the test pit, and the Sn values in the heavy mineral concentrates and the sieved till fractions show the achievement of a 80-130-fold enrichment by sluicing and panning.

The mineralogy of three heavy mineral samples from the pit and the compositions of their cassiterites are presented in Tables 4-5. The cassiterite grains in the samples are black, roundish chunks. The heavy minerals and the increased Nb and Ta concentrations in the cassiterite suggest that the source of the Sn and Ta heavy mineral anomalies should be a pegmatite, as was also found as a weathered bedrock product in the test pit. Nb and Ta minerals proper were not found in the heavy mineral samples, but the possibility of finding these in the pegmatite is fairly good.

The glaciogenic dispersion of cassiterite from its bedrock source can be followed 4 km in the direction of ice movement. Figures 7 and 8 give an impression of actively moving ice. The moraine hummocks of the area have been interpreted earlier as ablation dead-ice moraines (Virkkala, 1969), but the present dispersion patterns suggest that they belong to the landform class of active-ice hummocky moraines.

## 3. *Leteensuo Zn*

Leteensuo is situated 18 km northwest of the town of Hämeenlinna in southern Finland. The bedrock is composed of mica schists and basic and acid volcanics. The area has a gently undulating ground moraine landscape. The depressions between the hills are covered by silt and clay deposits. Leteensuo is located just inside the Third Salpausselkä ice-marginal formation, the last phase of ice flow having been from the west-northwest, towards the ice-marginal formation. The zinc concentrations of the heavy mineral samples are presented in Figure 10. The middle part of the figure features a zinc anomaly fan, with its diameter in the direction of the last ice flow. The visible length of the anomaly is 5.5 km, and its real length could be some 7 km, but the anomaly ends on the shore of Lake Lehijärvi and no samples have been taken from the bottom of the lake. Thus the anomaly also ends at the Third Salpausselkä complex. The proximal part of the anomaly

Table 4. Mineralogical composition of three heavy-mineral samples from Kanajärvi.

| 530 ppm Sn | 480 ppm Sn | 1310 ppm Sn |
|---|---|---|
| amphibole | almandite | almandite |
| chlorite | amphibole | amphibole |
| hematite | ilmenite | hematite |
| ilmenite | chlorite | chlorite |
| magnetite | magnetite | mica |
| rutile | mica | quartz |
| quartz | quartz | plagioclase |
| plagioclase | plagioclase | cassiterite |
| cassiterite | cassiterite | |

Table 5. Chemical composition of nine cassiterite grains in three heavy-mineral samples from Kanajärvi (in wt%).

| | 530 ppm Sn | | | 480 ppm Sn | | | 1310 ppm Sn | | |
|---|---|---|---|---|---|---|---|---|---|
| | 1 | 2 | 3 | 4 | 5 | 6 | 7 | 8 | 9 |
| MgO | 0.1 | 0.2 | 0.1 | 0.1 | 0.1 | 0.3 | 0.2 | 0.1 | 0.1 |
| $Al_2O_3$ | 0.1 | 0.1 | 0.1 | 0.1 | 0.1 | 0.1 | 0.1 | 0.1 | 0.1 |
| $TiO_2$ | 0.0 | 0.1 | 0.0 | 0.1 | 0.1 | 0.1 | 0.2 | 0.0 | 0.0 |
| $Fe_2O_3$ | 0.1 | 0.3 | 0.1 | 0.9 | 0.2 | 0.4 | 0.3 | 0.4 | 0.1 |
| MnO | 0.0 | 0.0 | 0.1 | 0.0 | 0.0 | 0.0 | 0.1 | 0.0 | 0.0 |
| $Nb_2O_5$ | 0.1 | 0.2 | 0.1 | 0.6 | 0.1 | 0.2 | 0.2 | 0.1 | 0.1 |
| $Ta_2O_5$ | 0.2 | 0.5 | 0.2 | 1.7 | 0.5 | 0.9 | 0.7 | 0.8 | 0.3 |
| $SnO_2$ | 99.8 | 98.9 | 99.9 | 94.3 | 97.2 | 97.0 | 95.5 | 95.3 | 95.6 |
| SUM | 100.4 | 100.3 | 100.6 | 97.8 | 98.3 | 99.0 | 97.3 | 96.8 | 96.3 |

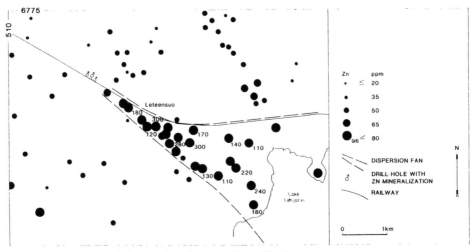

Figure 10. Zinc content in heavy mineral samples from till at Leteensuo.

is very narrow (200 m) and the distal part quite broad (2.5 km). Its maximum Zn content is 300 ppm (background 20 – 40 ppm).

The mode of occurrence of the zinc in the heavy mineral samples has not been studied, but low sulphur content and high iron and manganese values suggest that it is not present in the form of sulphides but rather in Fe-Mn hydro-oxides. A thin Zn mineralization drilled by Outokumpu Oy occurs 300 m northwest from the head of the anomaly and obviously acted as its source rock. The mineralization is located beneath a clay deposit, so that no till sampling could be performed directly above it.

### 4. *Hämeenlinna regional MgO, Ce*

Heavy mineral studies are also applicable to the deciphering of larger geological features. Figures 11 and 12 present the magnesium and cerium content of heavy mineral samples in the area west of the town of Hämeenlinna. Because the higher MgO content is caused by amphiboles and pyroxenes, MgO can be regarded as an indicator of basic and ultrabasic rocks. The cerium content is caused by monazite, so that Ce can be regarded as an indicator of acid granitic rocks. This area is very suitable for glacial transport studies, because the ice has moved across the strike of the bedrock (cf. Perttunen, 1977).

The magnesium anomalies form a zone roughly in the middle part of the area, following the strike of the bedrock (Fig. 11), the values rising at a distance of 3 – 5 km from the proximal contact of the basic volcanics in the direction of ice movement. It can also be said that up to this site granodioritic material has prevailed in the till matrix.

Thus the transport distance for the granodioritic material is 3 – 5 km. The magnesium anomalies extend in general 3 – 5 km from the distal contact of the basic volcanics. The only exception is the southeastern corner of Lake Takajärvi, where the anomaly extends 7 – 9 km from the distal contact.

We can see from Figure 12 that the southern granitic area is almost entirely anomalous with regard to Ce, which rises at a distance of 1 – 2 km from the proximal contact of the granite in the direction of ice movement, except in the southeastern corner of Lake Takajärvi, where the distance is 4 – 6 km. The cause of the longer transport distance southeast of Lake Takajärvi is probably that the area is a vast drumlin field (cf. Fig. 4), where the ice flowed more actively and transported the rock material farther than in adjoining areas. Lanthanum, thorium, phosphorus (monazite) and zirconium (zircon) also give a quite similar geochemical landscape to cerium. Since no sampling was carried out south of the granite area, it has not been possible to measure the transport distance for monazite there.

### 5. *Tervahaara W, Zn, S*

The study area is situated 37 km east of Oulu in northern Finland. The bedrock is composed of mica schists, black schists and basic volcanics. The landscape type is a ground moraine. The older ice flow was from the northwest and the younger one from the west-northwest. The amounts of scheelite grains in the heavy mineral samples are shown in Figure 13. In the eastern part of the area there occurs a narrow linear scheelite anomaly of length 500 m, the longitudinal axis of which parallels the younger direction of ice movement in the area. Study of the real shape of the anomaly was rendered difficult by the bog area and an esker chain north of the anomaly. The intensity of the scheelite

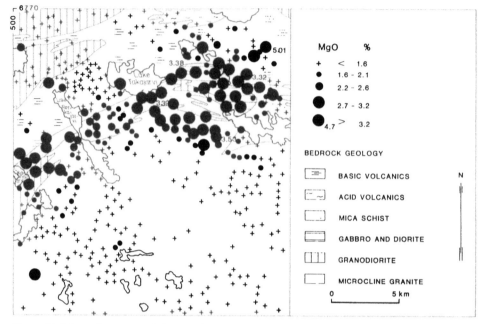

Figure 11. Magnesium content in heavy mineral samples from till in the Hämeenlinna region.

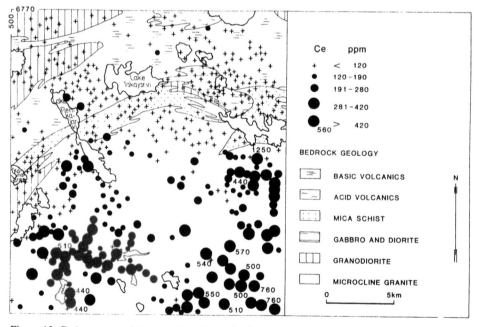

Figure 12. Cerium content in heavy mineral samples from till in the Hämeenlinna region.

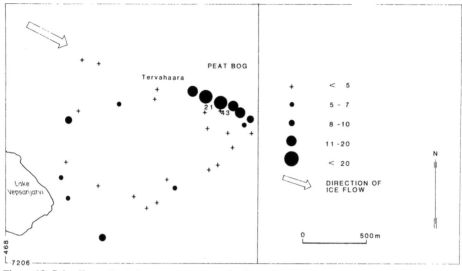

Figure 13. Scheelite grains in heavy mineral samples from till at Tervahaara.

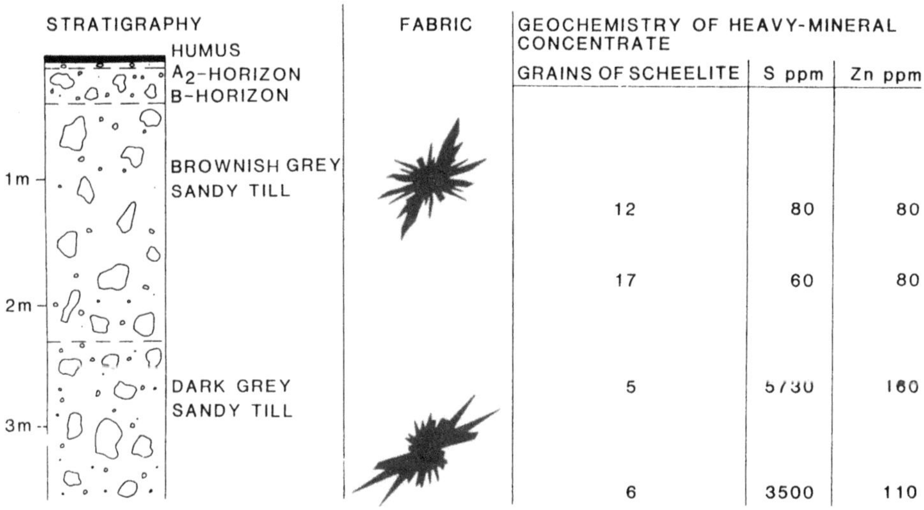

| STRATIGRAPHY | FABRIC | GRAINS OF SCHEELITE | S ppm | Zn ppm |
|---|---|---|---|---|
| | | GEOCHEMISTRY OF HEAVY-MINERAL CONCENTRATE | | |
| HUMUS, A₂-HORIZON, B-HORIZON | | | | |
| BROWNISH GREY SANDY TILL | | 12 | 80 | 80 |
| | | 17 | 60 | 80 |
| DARK GREY SANDY TILL | | 5 | 5730 | 160 |
| | | 6 | 3500 | 110 |

Figure 14. A test pit in the till at Tervahaara.

anomaly (max. 43 grains) is about the same as in Kanajärvi (example 2). A test pit was dug in the middle part of the anomaly with an excavator in order to determine the till characteristics (Fig. 14), but it was not possible to dig the pit any deeper than 3.5 m because of the groundwater. The stratigraphy of the pit is as follows. The upper bed of brownish grey sandy till with numerous rusty streaks is 2.25 m thick and has developed a distinct podsol profile. The lower bed is a dark grey sandy till reminiscent of the 'blue-grey till' described in many parts in Finland (Rainio and Lahermo, 1976, 1984).

The till beds differ only slightly in their fabric. The lower till has a more pronounced orientation of pebbles than the upper one, but the pebble orientations in both tills are transverse to the direction of ice movement. Thus the ice flow depositing the lower till must have been from a more northwesterly direction than that responsible for the upper till.

The sulphur and zinc concentrations in the lower till are considerably higher than those in the upper till. The reason could at least partly be that the upper till is more oxidized than the lower till, so that the sulphide minerals have been weathered away. The distribution of scheelite grains, which are much more numerous in the upper till than in the lower till, points to another explanation, however. Study of the bedrock source of scheelite grains is possible by means of an evaluation procedure called topographic control (Hirvas, 1980). This shows that the bedrock source of the scheelite anomaly in the upper till is to be found somewhere up-glacier from this site, on a rock-cored hill covered only by this upper till. If complex transport were involved, the scheelite anomaly would be less contrasted and less coherent.

The bedrock source of the scheelite anomaly has not been found because of difficulties in sampling procedures caused by glaciofluvial deposits.

## 6. *Pyyräselkä W*

The area is situated 45 km east of Oulu in northern Finland. The bedrock is composed of basic volcanics and mica schists and is covered by a rather thin sandy till. The type of landscape morphology is a ground and cover moraine with drumlins in places. The ice flow directions are the same as in the Tervahaara area (example 5). The numbers of scheelite grains in the heavy mineral samples are shown in Figure 15. The maximum numbers, 41 and 38 grains, are located in the marginal part of a vast magnetic anomaly and near an electromagnetic anomaly. The scheelite anomaly was studied further by till sampling with a percussion drill (Fig. 16). Samples were taken from as near the bedrock surface as possible. Tungsten was analyzed from the fine fraction of the samples (– 0.06 mm).

A quite coherent tungsten anomaly in the till fines occurs over the southern scheelite anomaly (41 grains) and west of it. The highest W content in the fine fraction is 55 ppm. Later drillings pierced several narrow quartz-carbonate veins in the bedrock which had a sparse scheelite dissemination, a source of the scheelite anomaly. The dispersion of scheelite in the fine fraction of the till can be traced 300 m from the mineralized veins in the direction of ice movement. Scheelite in the heavy mineral fraction of the till reaches farther from the source than that in the fine fraction, as far away as 2 km.

## 7. *Värriöjoki Cr*

The area is situated near the Finnish-Soviet border in eastern Lapland. The bedrock is composed of granitoids and an ultrabasic intrusion. The landscape is a gently undulating cover and ground moraine. In some places a thin preglacial weathering crust occurs beneath the till. The ice flow directions in glacial times fluctuated greatly, because the area is situated in the ice divide zone. The last deglacial ice flow was from the west-northwest (Hirvas et al., 1977).

The anomalous chromium values in the heavy mineral samples are presented in the

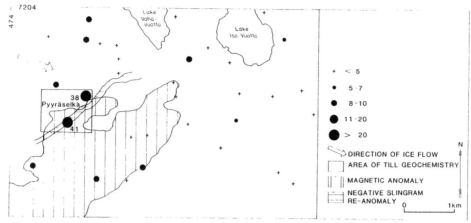

Figure 15. Scheelite grains in heavy mineral samples from till at Pyyräselkä.

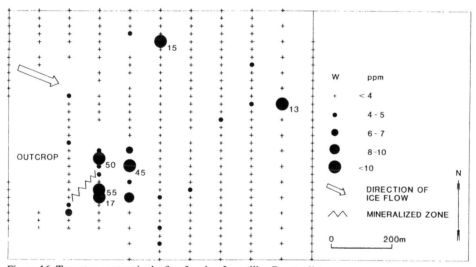

Figure 16. Tungsten content in the fine fraction from till at Pyyräselkä.

form of a contour map in Figure 17. The maximum Cr content is 7.16%. A ribbon-like chromium anomaly 24 km long and 5 – 6 km broad is oriented in accordance with the last direction of ice movement. The chromium occurs in the heavy mineral samples in the form of chromite. Further investigations led to an ultrabasic intrusion with sparse chromite dissemination being discovered in the proximal part of the anomaly (Vuollo, 1986).

Chromite is enriched to some extent in the lateritic weathering crust partly covering the intrusion as compared with the unweathered rock. The ultrabasic intrusion and its weathering crust together are the source rock for the chromite anomaly in the till, which

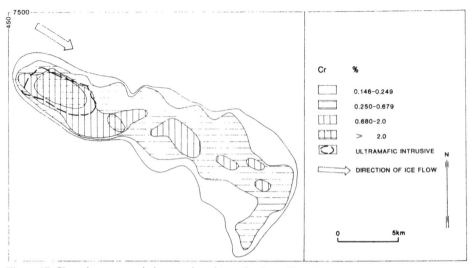

Figure 17. Chromium content in heavy mineral samples from till at Värriöjoki.

extends 17 km from the distal contact of the intrusion in the direction of ice movement. This is a rather long transport distance compared with the other examples quoted here (cf. Shilts, 1973; Ek and Elmlid, 1976), the main factors responsible for this having been the large surface area of the source rock (2 km × 6 km), the good resistance of chromite to chemical weathering and glacial milling and its good enrichment in panning.

CONCLUSIONS AND DISCUSSION

As the above examples show, the heavy minerals in till are very useful in till provenance and stratigraphical investigations and for prospecting with regard to the source rock of ore boulders. If the till of the area where an ore boulder has been found contains the same ore or indicator minerals in anomalous amounts as an ore boulder, the latter will in all probability be local in origin or the result of short-distance transport. If the till does not contain the same ore or indicator minerals as an ore boulder, the latter will most probably be a product of long-distance transport and more difficult to trace back to its source in the bedrock. In some cases an ore boulder can have been broken up and weathered, leaving no trace on the ground surface, although even then some ore minerals can have survived deeper in the till. Thus, although boulder hunting does not necessarily give an indication of the presence of an ore, this can be obtained through heavy mineral studies. This can be regarded as a rule. Heavy mineral studies succeed best when the minerals used are resistant ones, e.g. cassiterite, chromite or scheelite. Chromite is perhaps the most applicable mineral when long transport distances are to be investigated. The more easily weathering minerals such as sulphides, can also give good results, however. An excellent example of this is the Zn anomaly fan at Leteensuo (example 3).

The transport distance for heavy minerals, i.e. the distance from the source to the point where the minerals can be found in anomalous amounts, may be several kilometres

Table 6. Transport distances in the study areas on the basis of heavy-mineral data.

| Study area | Mineral | Landscape type | Transport distance (km) |
|---|---|---|---|
| Rakokivenmäki | cassiterite | cover moraine | 5 |
| Kanajärvi | cassiterite, scheelite | active-ice hummocky moraine | 4 |
| Leteensuo | sphalerite(?) | cover and ground moraine | 5.5 |
| Hämeenlinna regional | amphiboles, pyroxenes, monazite | cover moraine, active-ice hummocky moraine, drumlin | 3-5 7-9 |
| Pyyräselkä | scheelite | cover and ground moraine | 2 |
| Tervahaara | scheelite | cover and ground moraine | 0.5 |
| Värriöjoki | chromite | cover and ground moraine | 17 |

(Table 6), and is usually very much longer than that of geochemical anomalies in the fine till fraction. Thus heavy mineral anomalies can be found using a less dense sampling grid so that the method can be used to survey vast areas quite quickly. It could be inferred from the data in Table 6 that glacial transport in drumlin areas is somewhat more far-reaching than in cover moraine, ground moraine or active-ice hummocky moraine areas.

Till stratigraphy studies are necessary in heavy mineral investigations, and the concentration method used can be selected on the basis of demands concerning costs, time and accuracy. In the chemical analysis it is sensible to determine a great number of elements. Mineralogical investigations are rather slow to perform and should therefore be concentrated on samples of particular interest selected on the basis of chemical analyses.

ACKNOWLEDGEMENTS

Mrs. Kristiina Karjalainen has drawn the maps and diagrams. Mrs. Hilkka Määttä has copied the maps and diagrams for printing. Mrs. Raili Junnila has typed the manuscript. The English of the manuscript was checked by Mr. Malcolm Hicks, M.A. I express my sincere gratitude to all these people.

REFERENCES

Bates, P.L. and J.A. Jackson, 1980. Glossary of Geology. American Geological Institute, Falls Church, Virginia, 749 pp.
Brundin, N.H. and J. Bergström, 1977. Regional prospecting for ores based on heavy minerals in glacial till. Journal of Geochemical Exploration 7: 1-19.
Callahan, J.E. 1981. A regional stream sediment and heavy mineral concentrate survey, Churchill Falls, Labrador. Canadian Institute of Mining and Metallurgy Bulletin 74(829): 100-114.
Carroll, D. 1953. Weatherability of zircon. Journal of Sedimentary Petrology 23: 106-116.
Closs, L.G. and E.V. Sado, 1982. Orientation Overburden Geochemistry and Quaternary Geology Investigations of Carbonatite-Alkalic Complex in the Prairie Lake and Killala Lake Areas, District of Thunder Bay. Ontario Geological Survey, Study 23, 65 pp.

Day, S. and K. Fletcher, 1986. Particle size and abundance of gold in selected sediments, southern British Columbia, Canada. Journal of Geochemical Exploration 26: 203-214.

DiLabio, R.N.W. and W.W. Shilts, 1979. Composition and dispersal of debris by modern glaciers, Bylot Island, Canada. In: Ch. Schlüchter (ed), Moraines and Varves. A.A. Balkema, Rotterdam: 145-155.

Dreimanis, A. and G.H. Reavely, 1953. Differentiation of the lower and the upper till along the north shore of Lake Erie. Journal of Sedimentary Petrology 23: 238-259.

Dudykina, A.S., 1959. Parageneticheskie assotsiatsii elementov-primesei v kassiteritakh pazlichnykh geneticheskikh tipov olovorydnykh mestopozhdenii (in Russian). Akad. Nauk. SSSR. Voprosy geohimii 28: 111-121.

Dunlop, A.C. and W.T. Meyer, 1978. Detrital tin patterns in stream sediments and soils in Mid-Cornwall. Journal of Geochemical Exploration 10: 259-276.

Ek, J. and C.G. Elmlid, 1976. Tjärrovare – chromium in till. Journal of Geochemical Exploration 5: 349-364.

Evenson, E.B., T.A. Pasquini, R.A. Stewart and G. Stephens, 1979. Systematic provenance investigations in areas of alpine glaciation: applications to glacial geology and mineral exploration. In: Ch. Schlüchter (ed.), Moraines and Varves. A.A. Balkema, Rotterdam: 25-42.

Fletcher, W.K., 1981. Analytical Methods in Geochemical Prospecting. In: G.J.S. Govett (ed.), Handbook of Exploration Geochemistry. Elsevier, Amsterdam, 255 pp.

Hirvas, H., 1980. Moreenistratigrafiasta ja sen merkityksestä malminetsinnässä. Geologi 32(4): 33-37.

Hirvas, H., A. Alfthan, E. Pulkkinen, R. Puranen and R. Tynni, 1977. Raportti malminetsintää palvelevasta maaperätutkimuksesta Pohjois-Suomessa vuosina 1972-1976. Summary: A report on glacial drift investigations for ore prospecting purposes in northern Finland 1972-1976. Geological Survey of Finland, Report of Investigation, 19 54 pp.

Jones, M.P. and M.G. Fleming, 1965. Identification of Mineral Grains. Elsevier, 102 pp.

Kokkola, M. and E. Pehkonen, 1976. Kangaskylä – gold in till. Journal of Geochemical Exploration 5: 239-244.

Kuzin, M. and N. Egorov, 1976. Field Manual of Minerals. Mir Publishers, Moscow, 194 pp.

Lee, H.A., 1963. Glacial fans in till from the Kirkland Lake fault – a method of gold exploration. Geological Survey of Canada Paper 63-45: 1-36.

Lee, H.A., 1965. Investigation of eskers for mineral exploration. Geological Survey of Canada Paper 65-14: 1-17.

Lee, H.A., 1968. An Ontario kimberlite occurrence discovered by application of the glaciofocus method to a study of Munro esker. Geological Survey of Canada Paper 68-7: 1-3.

Lehmuspelto, P., 1976. Eurajoki – tin in till. Journal of Geochemical Exploration, 5: 218-221.

Lindmark, B., 1977. Till-sampling methods used in exploration for scheelite in Kaustinen, Finland. In: M.J. Jones (ed.), Prospecting in Areas of Glaciated Terrain 1977, Institution of Mining and Metallurgy, London: 45-48.

Mertie, J.B., 1954. The gold pan: a neglected geological tool. Economic Geology 49: 639-651.

Mertie, J.B., 1979. Monazite in Granitic Rocks of the Southeastern Atlantic States – an Example of the Use of Heavy Minerals in Geologic Exploration. U.S. Geological Survey, Professional Paper 1094, 79 pp.

Nikkarinen, M. and A. Björklund, 1976a. Emmes – the use of till in spodumene exploration. Journal of Geochemical Exploration 5: 212-218.

Nikkarinen, M. and A. Björklund, 1976b. Kaustinen – the use of till in tungsten prospecting. Journal of Geochemical Exploration 5: 247-248.

Ossenkopf, P., C. Erbe and P. Zurlo, 1979. Methodische Ergebnisse von Schlichprospektionen. Zeitschrift für angewandte Geologie 25: 522-527.

Perttunen, M., 1977. The lithologic relation between till and bedrock in the region of Hämeenlinna, southern Finland. Geological Survey of Finland, Bulletin 291, 68 pp.

Peuraniemi, V., 1981. Moreenin pintaosan raskasmineraaligeokemia tinamalmin etsinnässä Rakoki-venmäellä Etelä-Suomessa (in Finnish). In: P. Lindroos (ed.), Pintamoreenin merkitys malmilohkarekuljetuksessa. Geological Survey of Finland, Report of Investigation 5: 61-69.

Peuraniemi, V., 1982. Geochemistry of till and mode of occurrence of metals in some moraine types in Finland. Geological Survey of Finland, Bulletin 322, 75 pp.

Peuraniemi, V., 1987. Interpretation of heavy mineral geochemical results from till. Geological Survey of Finland, Special Paper 3: 169-179.

Peuraniemi, V., E. Mattila, J. Nuutilainen and H. Autio, 1984. Till and bedrock geochemistry in Sn exploration: a case study. Journal of Geochemical Exploration 21: 249-259.

Reeves, R.D. and R.R. Brooks, 1978. Trace Element Analysis of Geological Materials. John Wiley & Sons, New York, 421 pp.

Saarnisto, M. and E. Tamminen, 1985. Lapin kultaprojekti 1982-1985. English summary: Lapland gold research project 1982-1985. Geologi 37: 157-162.

Saarnisto, M. and E. Tamminen, 1987. Placer gold in Finnish Lapland. Geological Survey of Finland, Special Paper 3: 181-194.

Saxby, D. and K. Fletcher, 1986. The geometric mean concentration ratio (GMCR) as an estimator of hydraulic effects in geochemical data for elements dispersed as heavy minerals. Journal of Geochemical Exploration 26: 223-230.

Shilts, W.W., 1973. Glacial Dispersal of Rocks, Minerals, and Trace Elements in Wisconsinan Till, Southeastern Quebec, Canada. Geological Society of America. Memoir 136: 189-219.

Simonen, A., 1949. Kallioperäkartta-Pre-Quaternary rocks, Sheet 2131, Hämeenlinna. Geological map of Finland, 1:100 000.

Tencik, I. and P. Sponar, 1979. Some problems and perspectives of the heavy mineral prospection. Symposium of Methods of Geochemical Prospecting, Prague, Vol. I, I: 174-179.

Theobald, P.K. and C.E. Thompson, 1959. Geochemical prospecting with heavy-mineral concentrates used to locate a tungsten deposit. U.S. Geological Survey Circular 411: 13 pp.

Thompson, I.S., 1979. Till prospecting for sulphide ores in the Abitibi Clay Belt of Ontario. Canadian Institute of Mining and Metallurgy Bulletin 72(807): 65-72.

Toverud, Ö., 1979. Humus: a new sampling medium in geochemical prospecting for tungsten in Sweden. In: M.J. Jones (ed.), Prospecting in Areas of Glaciated Terrain 1979. Institute of Mining and Metallurgy, London: 74-79.

Toverud, Ö., 1982. Geochemical prospecting for tin in upper central Sweden using heavy-mineral concentrate of glacial till. In: P.H. Davenport (ed.), Prospecting in Areas of Glaciated Terrain 1982, Canadian Institute of Mining: 204-212.

Toverud, Ö., 1984. Dispersal of tungsten in glacial drift and humus in Bergslagen, southern central Sweden. Journal of Geochemical Exploration 21: 261-272.

Virkkala, K., 1969. Suomen geologinen kartta 1:100 000 Hämeenlinna, Maaperäkartan selitys. Explanation to the map of Quaternary deposits. Geologinen tutkimuslaitos, The Geological Survey of Finland, 69 pp.

von Philipsborn, H., 1967. Tafeln zum Bestimmen der Minerale nach äusseren Kennzeichen. Stuttgart, 267 pp.

Vuollo, J., 1986. Värriöjoen ultraemäksisen intruusion petrologia, mineralogia ja geokemia. Unpublished Master Thesis, University of Oulu, 108 pp.

Watters, R.A., 1983. Geochemical exploration for uranium and other metals in tropical and subtropical environments using heavy mineral concentrates. Journal of Geochemical Exploration 19: 103-124.

Zantop, H. and J. Nespereira, 1979. Heavy-mineral panning techniques in the exploration for tin and tungsten in northwestern Spain. In: Geochemical Exploration 1978, Golden, Colorado: 329-336.

# CHAPTER 11

# Geochemical-mineralogical profiles through fresh and weathered till*

W.W. SHILTS & I.M. KETTLES
*Geological Survey of Canada, Ottawa, Ontario*

## INTRODUCTION

To use glacial sediments effectively in mineral exploration, it is necessary to understand the causes of vertical compositional variability, whether they are sedimentologic, stratigraphic, or diagenetic. Many useful sedimentologic models are recorded in recent literature, particularly for till and related diamictons, and the importance of understanding contrasting compositions of different stratigraphic units is well recognized, if not always adequately dealt with in mineral exploration literature. Diagenetic alterations of drift, on the other hand, have received scant attention in the drift geochemical literature, except for discussions of the mineralogical and chemical alterations that take place in the true solum.

The primary objective of this paper is to document sub-solum diagenetic alteration of a thick till sheet, which contains abundant concentrations of labile minerals, some of which have economic significance. The alteration described takes place in the zone of oxidation, above the ground water table. Alteration is particularly easy to document in the many well-exposed till sections of the Appalachian region of southeastern Quebec; colour contrasts and chemical changes in various size and specific gravity fractions of till attest to the effects of post depositional weathering.

A secondary objective is to describe mineralogical and chemical partitioning among various grain sizes in till. Mineral phases most susceptible to weathering, carbonates and sulphides, are abundant in Appalachian tills and have long been known to concentrate in specific size fractions because of their varying resistance to glacial abrasion (Dreimanis and Vagners, 1971; Shilts, 1973, 1984).

To study mineralogical partitioning of carbonates and sulphides in Quebec till, several samples were divided into six size fractions from 6 mm to particles finer than 2 μm. Determining the size ranges within which sulphide and carbonate minerals are most highly concentrated allows selection of the best laboratory methods for effective routine analysis of weathered and unweathered samples.

A third objective is to compare carbonate contents of the various grain sizes using three different laboratory procedures – the Chittick method (Dreimanis, 1962), Leco Carbon Analyzer (Foscolos and Barefoot, 1970), and total dissolution. Glacial deposits are frequently analyzed for carbonate content using one or more of these techniques. Results

*Geological Survey of Canada, Contribution No. 31288.

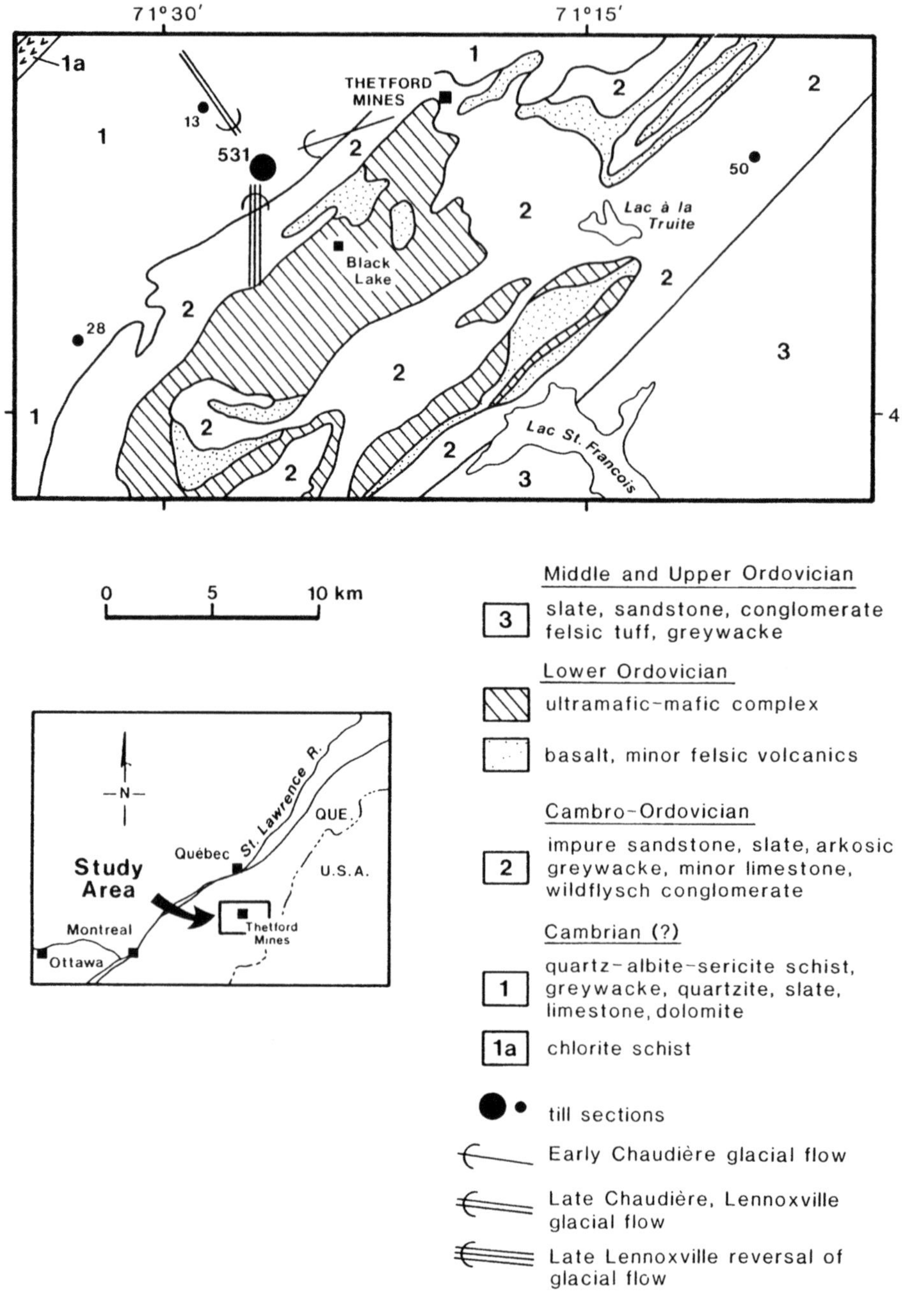

Figure 1. Location map and geology of the region around section 531.

of analyses obtained using the three different methods are compared and the methods evaluated.

In order to address these objectives, an exposure of thick Lennoxville Till (Shilts, 1981), designated 'section 531' and located near Thetford Mines, Quebec (Fig.1), was selected for detailed sampling and analyses. The section was deemed suitable because: 1) it is high and well exposed; 2) it is in an area where there is good knowledge of bedrock and glacial geology; 3) the till contains high concentrations of minerals that are sensitive indicators of weathering – pyrite cubes large enough to be seen by the naked eye and some carbonate minerals; and 4) it is located in an area where geochemical data from sample profiles on several other exposures are available for comparison.

Section 531 is located on a stream variously known as Nadeau Creek or White Stream. It is a natural stream bank 14 metres high. The upper 10 m is cut through till, and the lower 4 m is obscured by slump.

In 1977 fourteen samples of till were collected at 50 to 40 cm vertical spacings from near the top to a depth of 10 m. In 1985, a further 5 samples were collected at and above the lower boundary of oxidized till, and in 1987 a further 2 samples were collected at the sites of till fabric measurements depicted on Figure 2.

At the base of its exposure, the till is massive, slightly calcareous, hard, compact, dark blue grey, and stoney, with subequal amounts of sand, silt, and clay in its matrix. Abundant evidence of shearing as well as some sand lenses were noted at 5.5 metres and 6.3 metres depth, respectively. Individual cubes, twinned cubes, and fractured cubes of pyrite are a conspicuous component of the sand to pebble-size fraction.

*Chemical analyses*

The size fractions of till used for partitioning studies, and fine sand (.125-.25 mm) for heavy minerals were obtained by washing the bulk sample through a series of sieves of mesh sizes 38 μm, 44 μm, 63 μm, 125 μm, 250 μm, 2.0 mm and 6.0 mm. The silt plus clay (<63 μm) fraction was obtained by dry sieving a split of the bulk sample through a 250-mesh sieve.

To separate the clay (<2 μm) for trace element analyses, approximately 350 grams of bulk sample was disaggregated and suspended in 0.05 N sodium hexametaphosphate solution in a milkshake mixer. The material suspended in the mixer bucket was collected and centrifiged at an appropriate speed to settle particles >2 μm diameter. The superna-tant suspension was decanted and recentrifuged at high speed for a time appropriate to settle particles >0.3 μm. The supernatant produced from this run was discarded, and the settled material was retained for analyses.

Total sulphur content was determined by a combustion analysis method (ASTM E320-47) using the Leco sulphur analysis apparatus. Within the Leco apparatus, a titration vessel filled with HCl, HI, KI, and starch solution contains a small amount of

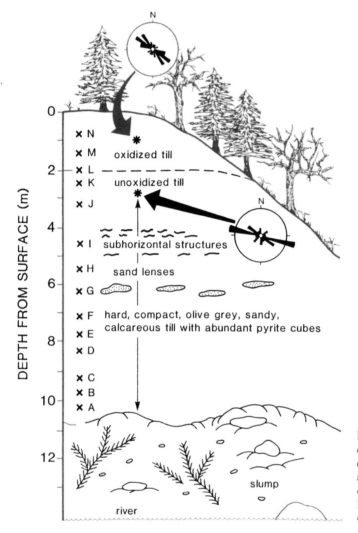

Figure 2. Schematic diagram of section 531; 2-dimensional fabric above and below base of zone of oxidation, upper part of Lennoxville Till; locations of 1977 samples.

$KIO_3$ added to release free $I_2$, thereby turning the starch solution blue. After a known weight of each sample fraction is combusted in oxygen, $SO_2$ produced from the burning of sulphur in the sample is conducted to the titration vessel, where it reacts with free iodine to form HI, which causes the starch solution to become colourless. The starch solution is titrated with $KIO_3$ until it returns to its original blue colour. Sulphur content of the sample is calculated on the basis of the amount of $KIO_3$ required to return the starch to its original colour.

Concentrations of Mn, Fe, Cu, Pb, Zn, Ni, Co, As, Sn, U, Ag, and Cd were determined for the clay-size fraction (<2 μm), the silt-clay fraction (<63 μm), and for sand-size heavy minerals (63-250 μm; s.g. >3.3). Pellets of dried clay and heavy mineral separates (minus magnetic minerals) were ground with an agate mortar and pestle to approximately <63 μm.

Analyses were carried out by Bondar-Clegg Co. Ltd. using standard atomic absorption spectroscopy, X-ray fluorescence, and colorimetric methods on samples leached in hot HCl-HNO3 for two hours. The accuracy of analysis is expected to be within ±15% of the stated values.

*Mineralogical analyses*

Heavy minerals were separated from a known weight of fine sand (63-250 μm) using methylene iodide (s.g. -3.3) and a separatory funnel, and their weight percent of the fine sand fraction calculated. Using an automagnet, magnetic minerals were removed and their weight percent of the heavy mineral fraction was determined. Some portions of the heavy mineral separates were used for trace element analyses, while the rest were mounted in araldite for point counting according to the method used by Paré (1982).

Determinations of percent calcite, dolomite, and total carbonate were made using the Chittick gasometric method described by Dreimanis (1962). The amount of $CO_2$ evolved from dissolution in HCl of carbonate in the <63 μm fraction of till was measured in a gasometric apparatus. Separate calculations of calcite and dolomite contents were made based on the difference in the rates of dissolution of these two minerals in HCl.

'Total' carbonate contents of the samples were also determined using a two-step process modified from the Leco method described by Foscolos and Barefoot (1970). A known weight of each sample was burned twice at high temperature – once untreated and once after treatment with HCl. Each time the amount of gas evolved was calculated from the amount of carbon detected in it. The gas evolved from the HCl-treated run is thought to be derived from non-carbonate (insoluable) carbon, such as graphite, Quaternary or modern plant detritus, and other organic compounds. The difference in carbon content between the treated and untreated samples was converted to % $CaCO_3$ equivalent, calculated as if all carbon was in $CO_2$ derived from thermal breakdown of calcite.

To calculate carbonate using the total dissolution method, a known weight of each sample fraction was leached with HCl to dissolve the carbonate minerals. The residue was then washed and filtered to remove CaCl precipitate before being dried and weighed. The difference in weight between the original sample and the dried residue was calculated as weight percent of the total fraction and was presumed to represent the weight of carbonate in the sample.

BEDROCK GEOLOGY

The Appalachian Highlands of the region around Thetford Mines form a gently rolling landscape with local relief rarely exceeding 200 metres. They are underlain mainly by steeply dipping, northeast-southwest striking metasedimentary and metavolcanic rocks, characterized in the vicinity of section 531 by quartz-albite-sericite schist containing abundant pyrite cubes as large as 1 cm in cross-section. A northeast-southwest striking belt of chlorite-albite-epidote-schists comprising metabasalts and rhyolites with known base-metal mineralization crops out 15 kilometres northwest of Section 531. A similar-striking belt of ophiolitic lithologies occurs 5 kilometers southeast of the section (Harron, 1976; Slivitzky and St. Julien, 1987).

GLACIAL GEOLOGY

The area around Thetford Mines has been glaciated at least three times, most recently by a glacier that flowed southeastward across the area from a dispersal centre on the Canadian Shield, then northward from a late glacial dispersal centre just SE of Thetford Mines (McDonald and Shilts, 1971; Lamarche, 1971; Shilts, 1981; Chauvin, 1979). Lennoxville Till, deposited during this glacial stade, is underlain elsewhere in the region by Chaudière Till, deposited by glaciers flowing first from Appalachian ice centres to the east, then southeastward from Laurentide centres on the Shield. Both of these glaciations are considered to be Wisconsinan; their ice-laid deposits in the Appalachians are commonly separated by the fine-grained glaciolacustrine deposits of Glacial Lake Gayhurst, a large proglacial lake dammed by glacier ice that stood against the northwest-facing Appalachian front and impounded water in the middle and upper reaches of the Chaudière and St. François and adjacent river valleys to altitudes as high as 427 metres, a.s.1.

This mid- to late-Wisconsinan sedimentary sequence is underlain in rare exposures by laminated glaciolacustrine silt-clay, fluvial gravels, and organic beds of the interstadial or interglacial Massawippi Formation. The Massawippi Formation is observed even more rarely to be underlain by Johnville Till, deposited by an earlier glacier flowing southeastward from the Canadian Shield.

Because section 531 is apparently cut entirely in Lennoxville Till, a brief discussion of flow trajectories immediately preceding and during its depositional cycle is appropriate. During the early part of the Chaudière glaciation, ice flow is thought to have been 250° to 260°. Because this trajectory is 20°-30° more northerly than the general NE-SW (230°) strike of structural elements in this part of the Appalachians, erratics transported west-southwestward from bedrock outcrops of units that are lithologically similar along strike may appear to have been transported northward from outcrops southwest of their true source (Fig. 1). This phenomenon is particularly troublesome along the ophiolite spite of the strong attenuation of erratic frequencies caused by reworking of the early Chaudière drift by southeastward flow during both the late Chaudière and Lennoxville glacial stades. Because section 531 is located a few kilometres northwest of the main ophiolite belt, any of the (very rare) physically and chemically distinctive ultramafic components found there could have been transported by southwestward flow as well as by northward flow (Fig. 1). McDonald (1967), Parent (1987), and Elson (1987) have all discussed this problem.

The separation of provenance indicators of northward flow from those of southwestward flow is particularly important for evaluating the petrology and geochemical composition of Lennoxville Till at section 531. The Lennoxville glacier, although exhibiting strong southeastward flow through most of the time during which it covered the Thetford Mines area, flowed northward from an ice divide located several kilometers southeast of section 531 during the latter part of its existence. This reversal of flow was in response to drawdown into a saddle created in the St. Lawrence valley by the upstream migration of a calving bay in the lower St. Lawrence estuary (Gadd, et al., 1972; Thomas, 1977; Shilts, 1981; Lortie, 1976). Striation evidence of late glacial northward flow is abundant on outcrops in the vicinity of section 531, but compositional evidence of the effects of northward flow is largely lacking at 531 and nearby sections. The few ultramafic erratics that have been found could as easily have been transported during the early phase of Chaudière glaciation from ophiolite outcrops northeast of the section.

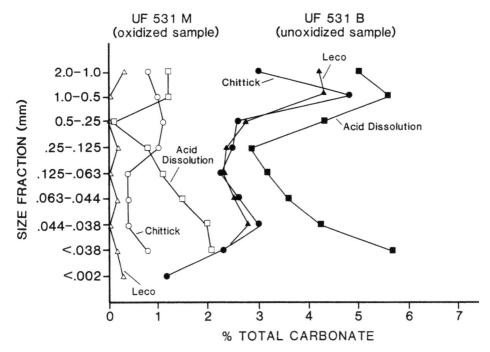

Figure 3. Comparison of three methods of determining carbonate concentrations in different size fractions of unoxidized (UF 531B) and oxidized (UF 531M) till, section 531.

Lennoxville Till at section 531 is thought to represent bands of basally transported debris, melted from the base of the glacier with little disturbance of the structure inherited from the ice, i.e. basal melt-out till. Fabrics measured 2 m and 4 m from the top of the section are strong and clearly reflect the southeastwardly flow of the Lennoxville glacier (Fig. 2).

At the top of nearby sections, one to two metres of a complex of crudely laminated sediments and till-like diamictons are often observed. These complex sediment packages of irregular thickness are considered to be supraglacial deposits, slumped from a retreating ice front or melted out of the upper portions of stagnating ice from which the underlying basal till was also melted. Because these surface sediments are often re-worked or disturbed by fluvial action, soil forming processes, tree root throw, frost processes, or agricultural activity, they are somewhat hard to discern, but the possibility of their occurrence at the top of a section should not be overlooked.

RESULTS

*Carbonate and sulphide-minerals*

*Comparison of 3 methods of carbonate analysis*
Leco, Chittick, and Total Dissolution methods of carbonate analysis were compared for various fractions of samples B to M. Calculated total carbonate concentrations were

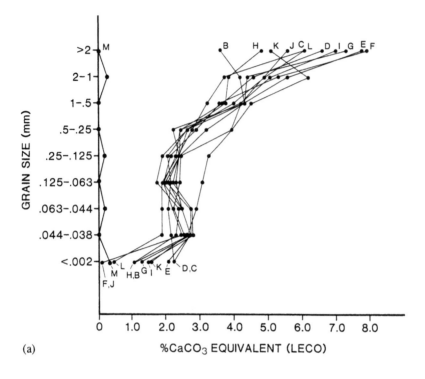

(a)

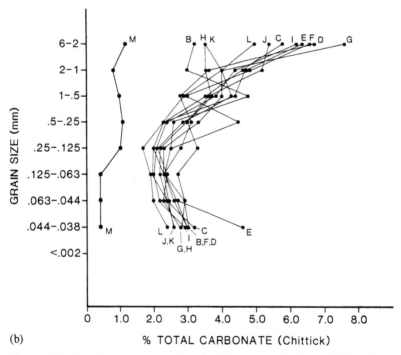

(b)

Figure 4. Total carbonate concentrations in different size fractions of till; a) determined by LECO method; b) determined by Chittick method.

generally similar for the Leco and Chittick methods, except in the granule (2-6 mm) fraction of some samples, but were consistently around 20% higher for the Total Dissolution method. Results of analyses for sample B, shown in Figure 3, are typical of those of other unoxidized samples collected at section 531. Sample M is representative of oxidized samples.

The comparative study of carbonate analyses shows total dissolution data to be unreliable, particularly when the sub-silt fraction is included in the fraction analyzed. In total dissolution, carbonate concentrations are calculated on the basis of total weight loss, which is assumed to represent destruction of carbonate minerals only. As well as dissolving carbonate, however, HCl also dissolves some forms of chlorite and other phyllosilicates, and Fe – Mn oxides. In oxidized, and/or leached samples of the normally low carbonate till derived from metamorphic terranes of the Appalachian Highlands, the quantity of non-carbonate minerals dissolved by HCl and removed in solution during washing increases the total weight loss of the sample, hence increasing apparent concentrations of 'carbonate' as shown by Figure 3, sample M. Weight loss associated with filtering to remove CaCl could be eliminated by allowing the HCl to evaporate, but weight loss calculations would still be affected because CaCl precipitate would remain in the residue.

The largest discrepancy between concentration levels obtained using total dissolution and the other two methods was found in the <38 μm fraction, which contains both the clay-sized phyllosilicates as well as secondary Fe/Mn oxide products of weathering. These phases, because of their solubility when present in fine sizes, readily react with HCl. Their dissolved components are removed in the filtering process, contributing to weight loss that enhances the apparent carbonate concentration (Fig. 3).

Although HCl is used to dissolve carbonate in the Chittick as well as the total dissolution method, the carbonate content calculated using the Chittick method is based on the amount of $CO_2$ evolved; $CO_2$ is produced only upon dissolution of carbonate carbon. Results obtained using the Chittick method were comparable to those obtained using the Leco method, where calculations are also based on the output of $CO_2$ gas. Of the two methods, however, the Leco is faster and more economical to use after the initial purchase of equipment. It is not known why the Chittick method yields indication of some carbonate when the Leco method indicates that virtually none is present in sample M (Fig. 3).

*Carbonate partitioning*
Carbonate concentrations in unoxidized samples vary markedly with grain size (Fig. 4), as would be anticipated from the work of Dreimanis and Vagners (1971). As they observed, carbonate concentration increases noticeably in the silt sizes, but is relatively more enriched than they noted in the fractions coarser than 0.5 mm. The reason for this is not presently known, but it must relate to greater proportions of non-calcareous debris in the monomineralic sizes (<0.05 mm), perhaps derived from easily eroded, non-calcareous terrain just up-ice from the section.

It is clear on Figure 4 that the variability of carbonate concentrations in the granule (>2 mm) fraction is greater than in finer grain sizes. This is probably because there are far fewer particles in this size range, and the compositional influence of a single particle can be proportionately greater. This is similar to the 'nugget effect' (Clifton et al., 1969) so familiar in gold exploration in glaciated terrain.

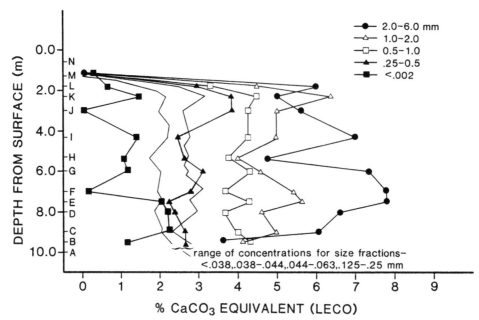

Figure 5. Vertical variations of equivalent CaCO₃ concentrations in various size fractions of till (LECO method).

*Carbonate leaching and provenance*

The carbonate content of unweathered Lennoxville and other tills of this region is consistently within the range of 2 to 10% and is frequently less than 4% (Shilts, 1978, 1981; S.J. Courtney, pers. comm., 1988). Although there is some limestone and dolomite within the generally non-calcareous formations of the region, most of the finer-grained carbonate minerals in the tills were probably derived from southeastwardly transported unmetamorphosed limestone and dolomite of the St. Lawrence Lowlands. One or two fragments that appear to be from the lowlands were found on section 531.

Within the oxidized zone at section 531, all size fractions of sample M, collected 1.5 metres from the top of the exposure, are carbonate-poor (Fig. 5). These laboratory data are borne out by field observations; even carbonate-rich granules or pebbles in till at this depth were punky remnants. In the next lower sample within the oxidized zone, sample L, concentrations of carbonate in all fractions except clay are diminshed but are not significantly lower than those of unweathered samples B-K.

*Partitioning of sulphur*

Sulphur concentrations in size fractions of till finer than 2 mm decrease systematically with decreasing grain size (Fig. 6a). The paucity of sulphur in the finer parts of the matrix suggests that only the fine sand-size and coarser pyrite fragments that are conspicuous components of the till were a major source of sulphur. Fragments of pyrite, a hard mineral with no cleavage, apparently are not ground easily to small particle sizes during glacial comminution.

Sulphur concentrations are most variable in the size fractions greater than 0.25 mm,

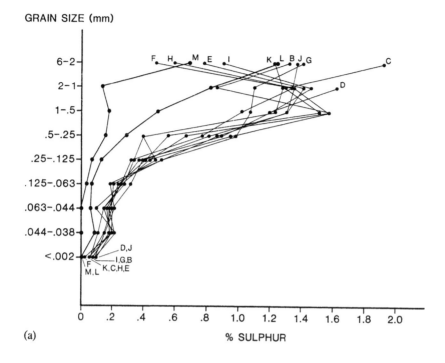

(a)

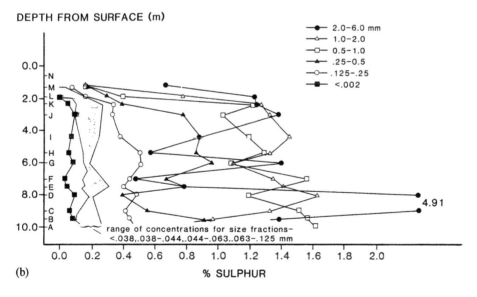

(b)

Figure 6. a) Partitioning of sulphur in various size fractions of till; b) vertical variations of sulphur concentration in various size fractions of till.

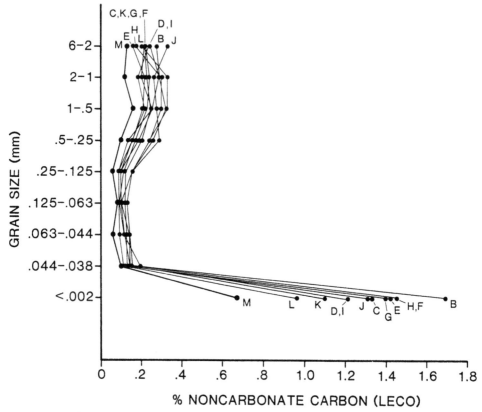

Figure 7. Partitioning of non-carbonate carbon in various size fractions of till.

particularly in the 2-6 mm range. The lack of homogeneity in the coarser fractions, similar to that found for carbonate, may also reflect the 'nugget effect'.

*Weathering profile of sulphur*

Sulphur concentrations are lower in brown to tan-coloured samples collected in the top 2 metres of the oxidation zone at the top of the exposure than in the lower, gray till samples (Fig. 6b). As would be expected if the sulphur is primarily derived from pyrite, its distribution over the sampling profile closely resembles that of Fe in heavy minerals (Fig. 8). When the samples were collected, it was observed that pyrite had been weathered to a limonitic mineral in the oxidized zone, but was fresh and usually untarnished in the gray, unoxidized till.

Small quantities of sulphur are present even in thoroughly oxidized sample M in size fractions coarser than silt. In tills from this part of the Appalachians, a large proportion of the sand and coarser size fractions are composed of rock fragments of slate and other fine-grained metasediments (Shilts, 1981, p. 18). These clasts are resistant to weathering except at their outer surfaces, relative to monomineralic grains. Any sulphide phases within the coarser, polymineralic clasts are protected from weathering but are released by

Table 1. Section 531: Heavy minerals (s.g. > 3.3).*

| Sample | Illmenite | Zircon | Chloritoid | Garnet | Epidote | Pyrite | Leucoxene | Goethite |
|--------|-----------|--------|------------|--------|---------|--------|-----------|----------|
| 531 N | 26 | 1 | 9 | 12 | 10 | 0 | 19 | 12 |
| 531 M | 28 | 1 | 11 | 12 | 11 | 2 | 19 | 5 |
| 531 L | 16 | 0 | 13 | 9 | 10 | 8 | 23 | 8 |
| 531 K | 15 | 2 | 14 | 11 | 6 | 22 | 14 | 4 |
| 531 J | 21 | 3 | 7 | 16 | 5 | 24 | 7 | 3 |
| 531 I | 22 | 0 | 7 | 11 | 5 | 28 | 9 | 5 |
| 531 H | 11 | 1 | 19 | 15 | 14 | 24 | 7 | 2 |
| 531 G | 17 | 0 | 8 | 12 | 6 | 31 | 12 | 1 |
| 531 F | 16 | 0 | 13 | 13 | 5 | 35 | 12 | 1 |
| 531 E | 17 | 0 | 13 | 17 | 7 | 24 | 14 | 1 |
| 531 D | 16 | 4 | 11 | 14 | 3 | 30 | 10 | 2 |
| 531 C | 16 | 0 | 16 | 16 | 13 | 22 | 5 | 0 |
| 531 B | 33 | 1 | 11 | 10 | 5 | 27 | 6 | 1 |
| 531 A | 12 | 2 | 14 | 17 | 11 | 26 | 8 | 3 |

* These percentages are calculated exclusive of pyroxenes and amphiboles, which make up 20 ±5% of this fraction, and exclusive of magnetic minerals which make up 15-20% by weight of the total heavy mineral fraction. Other minerals noted include rutile, unidentified opaque minerals, quartz, siderite, hematite, tourmaline, titanite, olivine and gahnite.

artificial crushing in the laboratory. Thus, preservation of this 'enclosed' sulphide component leads to high sulphur values in the coarser fraction of this till, whether the original sample was weathered or not.

*Partitioning of non-carbonate carbon*
Concentrations of non-carbonate carbon were determined as a by-product of the Leco method of carbonate analysis. Non-carbonate carbon comprised less than 0.1 to 0.3% of all size fractions except in clay (<2 μm), where concentrations were up to 15 times higher (Fig. 7).

In this part of the Appalachians, the major source of non-carbonate carbon is probably graphite or amorphous organic compounds in the black slates that typically form the bedrock of the region. Because of clast-on-bedrock or clast-on-clast abrasion associated with glacial erosion and transport, the organic components derived from these rocks have been comminuted preferentially to clay-size particles, both because of their original fine grain size and because their 'soft' physical properties.

*Profile analyses*

In addition to the carbonate, carbon, and sulphur data discussed above, vertical variation of four other distinctly different types of compositional data have been studied from section 531: 1) mineral percentages derived from counting of heavy mineral (s.g. >3.3) slides; 2) weight percentages of total heavy minerals and magnetic minerals of the fine sand fraction; 3) trace and minor element geochemistry of crushed heavy mineral (-magnetic minerals) separates from the fine sand fraction; and 4) trace and minor element geochemistry of the clay (<2 μm) and silt+clay (<63 μm) fractions. Geochemical

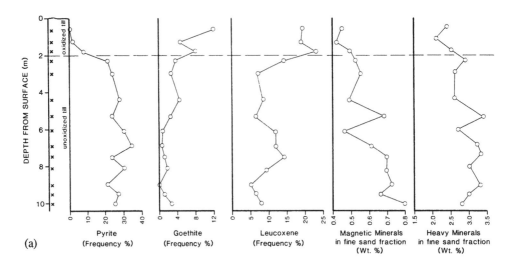

(a)

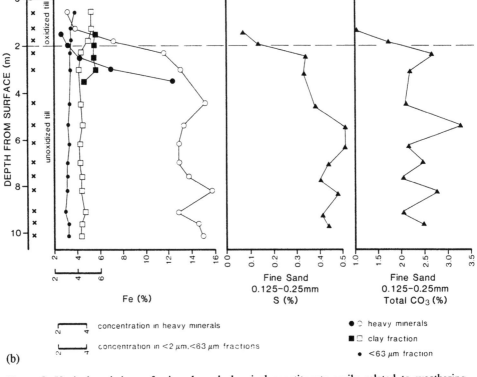

(b)

Figure 8. Vertical variations of minerals and chemical constituents easily related to weathering processes; a) profile of frequency % of pyrite and secondary oxides/hydroxides compared to weight % of magnetic and total heavy mineral fraction; b) Profile of Fe in clay, silt + clay, and heavy mineral fractions compared to S and CO$_3$ variations in fine sand. Larger solid symbols represent concentrations in 1985 samples.

profiles of clay and heavy mineral fractions were determined also for several nearby sections for comparison with section 531.

Point counting of non-magnetic heavy minerals (s.g. >3.3) was carried out to provide some basis for interpreting their chemical signature (Table 1). The heavy mineral grains listed in Table 1 are typical of non-magnetic mineral assemblages in Appalachian till, and account for 85-90% of all non-magnetic heavies in each sample. The remaining 10-15% of each sample includes rutile (2-4%), siderite (<1%), hematite (<1%), tourmaline (<1%), quartz (as a contaminant; <1%), unidentified black opaque minerals (1-7%), olivine (trace), titanite (trace), and gahnite (trace). In addition, clino and orthopyroxenes and hornblende are common in this specific gravity range, comprising, in aggregate, 20+ percent of the heavies, but because specific gravities of these minerals are so close to the 3.3 specific gravity of the heavy liquid (Methylene Iodide), very slight changes in laboratory procedures or specific gravity of the liquid could change their relative percentages radically. Consequently, the mineral frequencies were computed independently of these minerals.

Figure 8 summarizes vertical variations in mineralogy and in chemical parameters that can be related easily to observed mineralogical changes through the section. The most striking mineralogical variation in the section is the abrupt and total disappearance of pyrite above the base of the zone of oxidation. Since most sulphides have weathering characteristics similar to pyrite, we assume that they will be similarly depleted in oxidized samples. Thus, there must be a significant redistribution of chalcophile elements in the zone of oxidation, and it is likely that some cations are exported from the site in groundwater.

Because pyrite makes up such a large proportion of the heavy mineral fraction of till at this site, other, resistant heavy minerals show modest increase in concentration in oxidized till. This is a result of their making up a relatively larger part of the heavy mineral suite that is reduced in volume by 20 ±% through the removal of pyrite and associated sulphides. In contrast, goethite and leucoxene, secondary minerals formed as a result of alteration of unstable iron and titanium-bearing minerals, are enriched in the oxidized zone by a factor of about 2, suggesting that they were created by weathering processes. Pseudomorphs of goethite after pyrite are commonly found in physically undisturbed oxidized tills of the region.

The breakdown of pyrite is reflected by decreasing weight percentages of non-magnetic heavy minerals in the fine sand fraction. The increase of goethite and leucoxene somewhat offsets this decrease, but it is, nevertheless, well defined, indicating that this weathered till is depleted in heavy minerals.

*Magnetic fraction*
The content of magnetic minerals varies greatly over the sampling profile, although it generally decreases from bottom to top. This variation may reflect one or more of the following: 1) changes in ice flow direction, and, therefore, provenance during accretion of till beneath ice or stacking of debris in basal debris bands. In the latter case the debris bands would have been deposited without mixing or major disruption by basal meltout; 2) the variation may be due to measurement errors, always a problem with such small samples; 3) variation may be a combination of 1 and 2, caused by inconsistent separation of silicate minerals (particularly serpentine) which contain magnetic inclusions, perturbing their true specific gravity and making them susceptible to removal with a hand

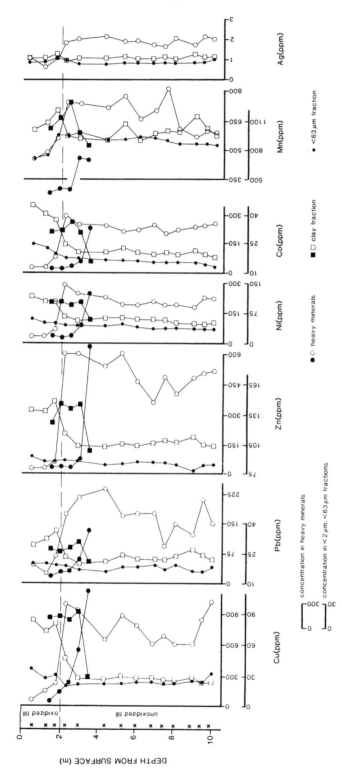

Figure 9. Vertical variations of trace elements, in clay (<2 μm), silt and clay (<63 μm), and non-magnetic heavy mineral (s.g. >3.3) fractions. Includes analyses of clay and heavy minerals from sample suite collected in 1985 (larger solid symbols).

magnet. Because magnetic minerals are most commonly dominated by resistate phases, it would be expected that they would be relatively enriched in thoroughly weathered samples. However, with the exception of the uppermost sample of section 531, the content of magnetic minerals in oxidized till was the same or lower than that in unweathered till below. This may reflect the breakdown of pyrrhotite, a magnetic sulphide more easily destroyed during weathering than magnetite. Grains thought to be pyrrhotite were found in magnetic separates from unweathered samples.

*Geochemical profiles*
Figures 8 and 9 depict the vertical trends of metal concentration for the sand sized heavy mineral (s.g. >3.3; minus magnetic minerals), silt plus clay-sized (<63 µm), and clay-sized (<2 µm) fractions of till from section 531. Two suites of samples were analyzed, a complete vertical sequence collected in autumn of 1977 and a vertical suite, collected in autumn of 1985 upward from the uppermost unoxidized till, through oxidized till into the true solum. Because no clear vertical datum was established after the first sampling was done, the vertical position of the second sample suit relative to the first was unknown. Therefore, the second sample suite, although reflecting almost exactly the compositional trends established from the first suite, is thought to be offset approximately 1.5 m. below the first – that is, the uppermost gray (unoxidized) till sample was collected at a depth measured as 3.5 m from the surface, whereas the uppermost unoxidized sample collected in 1977 was estimated to be only 2 m below the surface.

If the trace element profiles of the 1985 samples are raised 1.5 m on Figure 7, they correspond very closely to the profiles calculated from the 1977 samples, with one or two exceptions. The proven reproduceability of the trace element profiles is particularly important for several reasons: 1) the close correspondence of the curves confirms that the mineralogical changes inferred from the geochemistry are valid; 2) the analytical techniques and sampling and sample processing techniques are appropriate for characterizing the redistribution of cations among mineral phases as a result of weathering, as suggested by Shilts (1975); 3) changes in analytical equipment and procedures in 1979 at the commercial laboratory (Bondar-Clegg Co. Ltd.) that carried out the analyses have affected neither the absolute nor the comparative values for the elements analyzed, with the exception of Mn, discussed below; and 4) the ambient soil moisture (climatic) conditions at the time of sampling have had little effect on the absolute and relative concentrations of cations measured from the clay-sized and heavy mineral fractions of till. All in all, this comparative study gives us confidence that even the more subtle characteristics of the trace element profiles are real and that comparable profiles from sections in the region are equally valid and reproduceable.

One of the elements reanalyzed shows serious discrepancies in absolute concentrations. Although the shapes of the profiles are similar, Mn in the heavy mineral fractions yielded only about ½ of the concentration that was reported in 1977. The clay fractions from the two years yield comparable absolute values. The reason for this discrepancy is not known, but the magnitude and consistent relative trend of the error suggest that the analytical procedures are at fault. Because the original analyses were done over ten years ago, it is not possible to account for the error.

Figure 8b shows the geochemical expression of the destruction of sulphide phases so graphically shown by Tables 1 and 2 and by Figure 8a. Total iron, which is a major component in many heavy mineral phases, is reduced from levels over 13% in pyrite-rich

Table 2. Detailed non-magnetic heavy mineral (s.g. >3.3) analysis of oxidized and unoxidized till from till fabric localities*, section 531.

|  | Unoxidized % | Oxidized % |
|---|---|---|
| Pyrite | 17.0 | 0.0** |
| Goethite | 1.7 | 4.0** |
| Leucoxene | 2.7 | 4.0** |
| Garnet | 6.7 | 8 |
| Epidote | 7.2 | 7.5 |
| Orthopyroxene | 13.5 | 10.2 |
|   hypersthene | not determined | (9) |
|   bronzite | not determined | (1) |
| Clinopyroxene | 1.7 | 1.7 |
| Hornblende | 2.7 | 2.5 |
| Ilmenite | 20.0 | 23.7 |
| Chloritoid | 11.0 | 11.0 |
| Rutile | 5.0 | 8.0 |
| Hematite | 2.5 | 3.5 |
| Zircon | 1.5 | 2.7 |
| Siderite | 0.2 | 1.9 |
| Black Opaques | 2.0 | 6.0 |
| Marcasite | 0.2 | 0.0** |
| Quartz | 1.2 | 3.0 |
| Unknown | 1.0 | 1.2 |
| Total | 97.8 | 98.9 |

* See Figure 2
** Minerals altered or created by weathering

unweathered samples to 3-8% in weathered samples. Sulphur in the total fine sand fraction (0.125-0.250 mm) shows a decline in oxidized samples similar in pattern to that for iron, clearly reflecting the breakdown of sulphides.

Carbonate concentrations, while subject to considerable variation, probably due to analytical inconsistencies and possibly to 'nugget effect' problems, decline in the zone of oxidation. This indicates that the more common carbonate minerals are leached to the depth of oxidation in these carbonate-poor tills. However, the presence of siderite in oxidized samples, confirmed by X-ray diffraction analysis of several grains, indicates that not all carbonates are destroyed in this zone.

Figure 9 shows the relationship between trace and minor element concentrations in the clay, silt + clay, and non-magnetic heavy mineral fractions for chalcophile elements, Cu, Pb, Zn, Ni, Co, Ag, and for Mn. The similarities of the profiles of these elements to those profiles reflecting the visible evidence of weathering, colour change in the field and pyrite depletion or replacement, suggest that the elements are largely present in similarly labile minerals, probably sulphides. Although no sulphide minerals other than pyrite were identified by visual means at section 531, they have been found by scanning electron microscope with X-ray dispersive capability in heavy mineral separates from other sections, notably section 50 (Fig. 11c).

Unlike the heavy minerals, changes in the mineralogical composition of the till that could account for the increase in trace element levels in the clay-size and silt + clay size

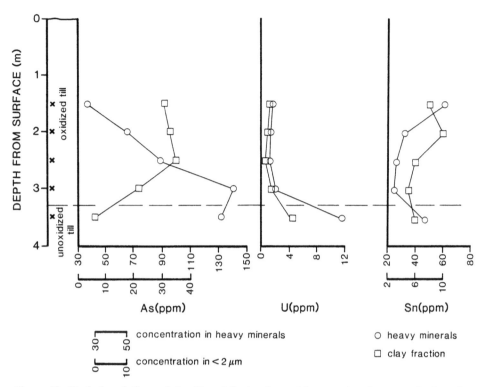

Figure 10. Vertical variations of As, U, and Sn in clay and heavy minerals across the boundary separating unoxidized and oxidized till; samples collected in 1985.

fractions of weathered till are not so easily observed. Although much of the metal leached from sand-size heavy minerals probably is transported from the site in solution in ground water, some is thought to be adsorbed onto phyllosilicate minerals and/or into aggregates or crystal structures of secondary oxides and hydroxides. It is also possible that aggregates of goethite and limonite, weathering products of sulphides and other minerals, are present in this fraction because the soft, crumbly structure of aggregates of these minerals is easily disaggregated to clay sizes during sample preparation in water.

On Figures 8 and 9, Fe, Co, Ni, Zn, Cu, and Pb show marked increases in concentration in clay and less sharp increases in silt and clay* above the base of the zone of oxidation, underlining the possible difference in background concentrations of these metals between oxidized and unoxidized samples of the same provenance.

Three additional elements were analyzed in heavy mineral and clay separates from

*Here, as for till throughout the region, most metal in the silt + clay fraction is derived from clay-size particles (Shilts, 1973; 1984). Thus, the concentration of metal in the silt + clay fraction is diluted by metal-poor silt. If the proportion of silt does not change significantly from sample to sample, as seems to be the case here, the compositional trends will be similar to those derived from clay analyses, but profile deflections and absolute values will be considerably dampened, as is clearly shown by Figures 8 and 9.

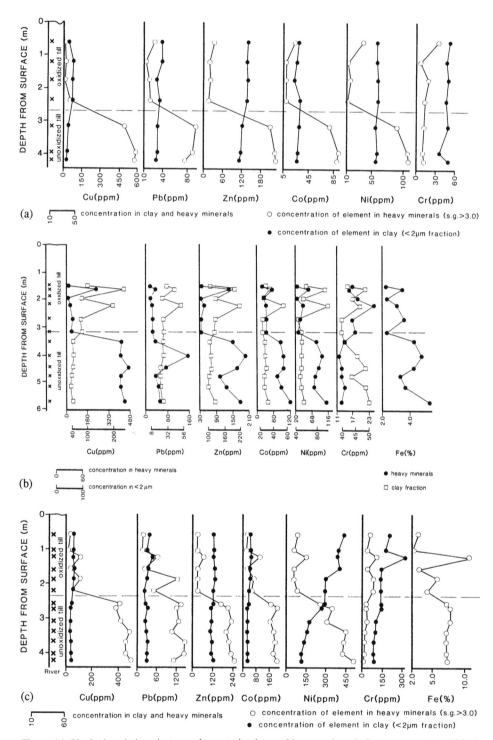

Figure 11. Vertical variations in trace elements in clay and heavy minerals from near section 531: a) section 13, Ruisseau Gregoire; b) section 28, White River; c) section 50, Petit Lac.

1985 samples collected from the upper part of section 531 (Fig. 10). Arsenic, a chalcophile element, shows the same pattern of decline in the heavy mineral fraction of oxidized till as other chalcophile elements, but the pattern of decrease is much less abrupt at the lower oxidation boundary. Similar to Fe, As increases in clay in the oxidized zone.

Uranium shows a decline in both heavy mineral and clay fractions from unweathered to weathered samples. The pattern of decline is not similar to those of chalcophile elements and the absolute concentrations of the two fractions are almost identical in the oxidized zone.

Tin in heavy minerals displays first a decrease, then an increase in concentration upward through the weathered zone. Tin in the clay-size fraction shows no strong trend. Because there are no specific local sources known for tin and uranium, it is not possible to speculate on whether these trends would be typical for other till, or not. Uranium-bearing erratics are known to occur in association with sulphide mineralization typical of the greenstones that crop out northwest of section 531 (unit 1a, Fig. 1).

*Other sections*
Trace element profiles for clay and non-magnetic heavy minerals are shown for three nearby till sections for comparison with results from section 531 (Fig. 11). Section 13, located a few km up-ice from section 531, has similar trace-element profiles for the heavy mineral suites, although the absolute concentrations of metal are strikingly lower than in section 531. Also, the marked enrichment in most metals of the clay fraction near the top of 531 is not reflected by this fraction at section 13. Chromium, presumably from non-magnetic chromite, was analyzed in section 13 and shows, as expected, no decline in the weathered zone.

Section 28 (Fig. 11b) has geochemical profiles that are similar to those from section 531, except for the lack of an apparent weathering decline for Pb. The shape of the Pb, Zn, and Fe curves, which all show a deflection with maximum metal enrichment at 4 m depth, suggests that some sedimentological or stratigraphic factor changed during deposition and/or entrainment of the debris in that part of the section. The late glacial reversal of flow (northward) may be responsible, but no obvious physical feature of the till corresponds with this break. As in section 531, the clay-sized fraction in the zone of oxidation is enriched in Cu, Pb, Zn, Co, and Ni. Cr is enriched in the heavy mineral fraction because of the stability of its presumed host mineral, chromite, with respect to the sulphide phases.

Section 50 (Fig. 11c) has been discussed before because of the sharp upward increase in nickel in the clay fraction (Shilts, 1976; Rencz and Shilts, 1980). The reason for the increase is the late glacial shift in ice flow from the main southeastward flow which brought little or no ultramafic debris across this site. During the latter part of the Lennoxville glaciation, entrained debris being transported southeastward from the massive ultramafic outcrops near Thetford Mines was redirected northward to section 50. Ni in the clay-sized fraction is contained in relatively resistate silicate mineral phases of ultramafic provenance, whereas Ni in the heavy mineral fraction is in labile sulphides derived from a presently unknown but non-ultramafic source. The nickel sulphides decline in the weathering zone just as they do at section 531.

Other chalcophile elements in non-magnetic heavies from section 50 behave similarly to the other nearby sections, but like those at section 13, clays show no near surface enrichment.

Table 3. Comparison of trace elements in non-magnetic heavy minerals (s.g.> 3.3) from till and nearby stream sediment.

White Stream

| Element | Unoxidized till | Stream sediment 200 m upstream | at section | 200 m downstream | Ratio of average stream concentration to till concentration |
|---|---|---|---|---|---|
| Cu (ppm) | 1100 | 244 | 248 | 328 | .248 |
| Pb (ppm) | 135 | 40 | 42 | 76 | .390 |
| Zn (ppm) | 640 | 88 | 71 | 80 | .124 |
| Mo (ppm) | 4 | 4 | 2 | 4 | |
| Co (ppm) | 220 | 104 | 125 | 134 | .550 |
| Ni (ppm) | 240 | 97 | 109 | 144 | .486 |
| Cr (ppm) | 14 | 12 | 10 | 14 | |
| Mn (ppm) | 440 | 720 | 760 | 680 | |
| Cd (ppm) | 2.8 | 0.2 | <0.2 | 0.2 | |
| Ag (ppm) | 2.0 | <0.1 | <0.1 | 0.2 | |
| Fe (%) | 12.5 | 11.6 | 13.0 | 16.8 | 1.104 |
| V (ppm) | – | 16 | 16 | 20 | |
| As (ppm) | 132 | 178 | 320 | 238 | 1.859 |

Bagot Brook

| Element | Unoxidized till | Stream sediment 200 m upstream | at section | 200 m downstream | Ratio of average stream concentration to till concentration |
|---|---|---|---|---|---|
| Cu (ppm) | 890 | 108 | 112 | 120 | .127 |
| Pb (ppm) | 66 | 14 | 16 | 12 | .212 |
| Zn (ppm) | 440 | 32 | 46 | 48 | .095 |
| Mo (ppm) | 5 | 2 | 2 | 1 | |
| Co (ppm) | 130 | 18 | 26 | 24 | .174 |
| Ni (ppm) | 217 | 21 | 24 | 33 | .120 |
| Cr (ppm) | 10 | 8 | 8 | 8 | |
| Mn (ppm) | 520 | 440 | 560 | 520 | |
| Cd (ppm) | 1.0 | 0.2 | <0.2 | <0.2 | |
| Ag (ppm) | 0.8 | <0.1 | 0.2 | <0.1 | |
| Fe (%) | 8.3 | 2.4 | 3.4 | 3.6 | .377 |
| V (ppm) | 16 | 10 | 12 | 12 | |
| As (ppm) | 300 | 60 | 71 | 85 | .240 |

*Trace elements in heavy minerals from till and stream sediments*

As part of the 1985 resampling programme, three stream sediment samples were collected upstream, at, and downstream from section 531. For comparison, three additional stream sediment samples were collected around an exposure of unoxidized till in the stream bed of Bagot Brook, about 5 km northwest of section 531. Pyrite placers were observed in the stream bed at this till outcrop in 1976, but were not evident in 1985. All stream sediments were fine to coarse sand collected from active longitudinal bars in the stream channels. Heavy minerals were separated from these samples and analyzed for trace and minor elements in order to compare chemistry of the source and derived sediments (first and third order derivative sediments of Shilts, 1976).

Table 3, in which chemistry of these sediment facies is compared, clearly shows several aspects of their relationship; 1) There is no correspondence between absolute concentrations of chalcophile elements in heavy minerals from till and stream sediment derived from it. Elements commonly found in more resistate minerals, however, generally have similar concentrations in both facies; 2) The ratio of metal in till to metal in stream sediments is not similar at the two sites. In fact, at section 531, the As and Fe concentrations in stream sediments actually exceed these elements in till. Co, Ni, and Cu are also very enriched in stream sediments relative to the same elements at Bagot Brook. Some tarnished pyrite was observed in stream sediments of section 531, but not in Bagot Brook. It can be concluded that the enhanced metal levels in section 531 stream sediments, relative to those at Bagot Brook, are related to presence of some sulphide washed from the till banks (Cu) and to formation of heavy secondary oxides, (goethite, etc.) which have scavenged metal released by sulphide weathering in the fluvial environment. The greater concentration in Fe and As in stream sediment heavies compared to till heavies at section 531 support these inferences. However, whatever the processes responsible for concentrating metal in stream sediment heavies at section 531, it is not taking place in weathered till. As shown elsewhere in this paper, the chalcophile elements are universally depleted to very low levels in heavy minerals from oxidized till at section 531 and elsewhere, in spite of the concentration of secondary iron oxides in the sand fraction (Table 1, 2).

DISCUSSION

From the data presented above, it can be seen that causes for vertical compositional variations in a till sheet can be complex, but in this case are largely related to weathering, with less common variations being related to provenance changes. It is extremely important to ascertain to what extent one or the other cause is responsible for vertical variation, first because weathered samples may not accurately reflect provenace, regardless of the fraction analyzed, and secondly because the abrupt lower boundary marking the compositional and physical effects of weathering may be mistaken for a contact between two different stratigraphic units or sedimentary facies, an error possibly made already in this region (Chauvin, 1979).

*Vertical variation due to weathering*

The major cause of vertical variation in till composition in section 531 and nearby sections is weathering. Signs of change can be detected visually at greater depths from the surface than they can from examination of geochemical data. Upward from a depth of 3.0 metres, yellow-brown staining, presumably from oxidation, was observed along joints in the till, though at this depth pyrite cubes are still 'fresh' and unweathered. With decreasing depth from the surface the intensity of oxidation increases, and till becomes yellow-brown in colour, soft to the touch, with limonitic spots or pseudomorphs which mark the position of former cubes or fragments pf pyrite. The colour change is largely caused by staining by precipitates of secondary iron oxides/hydroxides formed as a result of oxidation of pyrite. In olive grey tills characteristic of this region, any colour change due to oxidation provides a strong contrast to the original colour of unoxidized sediment and is therefore particularly eyecatching. Just below the surface of the section remnants of ultramafic and carbonate boulders and pebbles, altered or destroyed by weathering, are found.

Chemical alteration not visible to the eye is reflected in the results of laboratory analyses of samples. In the sand-sized part of the matrix, carbonate minerals, pyrite, sulphur, and trace elements in heavy minerals are depleted to depths of over 2.0 metres. The opposite occurs in the clay-size fraction where the trace element concentration tends to increase.

The correspondence of depths of carbonate leaching and oxidation at section 531 is similar to weathering patterns observed in southwestern New Brunswick, a geologically similar part of the Appalachian Highlands (Kettles and Wyatt, 1985). In contrast, shallow samples collected in moderately oxidized till lying on varied bedrock of the eastern part of the Precambrian Grenville structural province (Frontenac Arch) in central Ontario appear to be unleached and to contain substantial quantities of carbonate minerals (up to 30%) derived from Grenville marbles and Paleozoic limestones of the flanking St. Lawrence Lowlands (Kettles and Shilts, 1983).

The difference in depth of leaching and oxidation between the tills of the Quebec Appalachian region and those of the Frontenac Arch is possibly partially related to the greater abundance of pyrite and other sulphide minerals in tills of the former. Weathering of pyrite, a conspicuous component of the tills near Thetford Mines (though not of those in New Brunswick), involves the oxidation of iron and subsequent production of sulphuric acid. If not diluted by vigorous groundwater flow, any newly formed sulphuric acid reacts again with more pyrite or other sulphide minerals, and the process is accelerated (Levinson, 1980, p. 75-78). During this process, the sulphuric acid generated also reacts with carbonate minerals, enhancing their rate of dissolution. As well as having fewer sulphide minerals, the high concentrations of carbonate in Frontenac Arch tills suggests that they are better able to buffer soil acids and slow the leaching process. In New York, Merritt and Muller (1959) found that there are critical levels of carbonate concentration in till, above which the buffering capacity of till is disproportionately increased.

*Vertical variations due to provenance*

Striae measurements in this part of the Quebec Appalachians indicate ice flow during the late Wisconsin was southeastward and later northward (Lamarche, 1971; Shilts, 1981). The till at section 531 is thought to be deposited from southeastward flowing ice. Erratics (>6 mm) in the till consist of mica schist (95%), pyrite (2-5%), vein quartz (1-2%), magnetite-bearing greenstone (<1%), traces of Precambrian crystalline gneiss and very rarely fragments of ultramafic rock. The major source area of the rare ultramafic erratics is the northeast-southwest trending belt of ophiolitic lithologies located only 3 km to the south (Fig. 1). The position of this belt of rock with respect to former ice flow trajectories in the region, discussed previously, means that these erratics could have been deposited during either an earlier episode of southwestward flow or during the latest flow toward the north. Fabric, the extreme paucity of ultramafic erratics, and the lack of Co, Ni, and Cr enrichment in the till of section 531 suggest that it was deposited predominantly from southeastward flowing ice. Northward flowing ice did not disturb this site and left no detectable deposits over the till of NW provenance.

Nothwithstanding the sampling results from Lennoxville Till at sections 531 and 13, which may owe their lack of reworking or subsequent sediment deposition to proximity to the zone of low-velocity ice flow that must have characterized the Quebec Ice Divide, sections 28 and 50, located farther from the Divide, show signs of vertical compositional change related solely to shifts in provenance. It is suggested that any drift prospecting sampling program be preceded by a profile sampling study to clarify the regional implications and relationships between vertical variations caused by weathering and those caused by provenance shifts.

*Till classes*

Because of the striking colour contrast between oxidized and unoxidized samples in this part of the Appalachians, a simple system was devised to guide both experienced and inexperienced samplers and to simplify interpretation of analytical results. Regional till samples have been described according to a 'class' scheme where 'Class I' tills are unoxidized. They are olive grey, contain visible, fresh sulphide grains, and have no leached carbonate-rich clasts. Till exposures that yield 'Class I' samples often have fissility or subhorizontal bedding indicative of basal meltout. 'Class I' samples are found at depth (>2 m) in natural and man-made exposures or at shallow depth (<1 m) in reducing environments, such as beneath organic-rich flood plain sediments.

Class II tills are olive gray, transitional to tan hues and have sulphide grains in the beginning or final stages of destruction by oxidation. They are fairly uncommon, but include tills in the initial stages of oxidation or formerly oxidized till brought into a reducing environment and gleyed.

'Class III' tills are thoroughly oxidized, generally to a tan colour, but are otherwise physically similar to Class I tills. In this region, intact pseudomorphs of goethite after pyrite can be seen and carbonate-rich pebbles, where present, are reduced to punky remnants. South and east of section 531, down-ice from the Devonian granodioritic plutons of the New Hampshire Plutonic Series, clasts of granodiorite are decomposed to sandy pockets with original granitic structure intact.

'Class IV' till samples include thoroughly oxidized till that has been disturbed near the surface. Class IV till often has subhorizontal laminations caused by postdepositional downslope movement. This class is also characterized by mottled colour and inclusions of organic or sorted sediment; the sediment also may be disturbed by tree root throw, frost processes, anthropogenic activities, etc. Considerable secondary, low-temperature chemical alteration of all mineral fractions can be expected in these samples.

Section 531 and nearby sections demonstrate graphically the compositional significance of the Class I to III transition. In natural sections cut by streams, the depths at which the transition takes place are probably greater than would be expected away from streams because the water table descends as it approaches these abrupt, high channel walls.

CONCLUSIONS

The principal problem in interpreting geochemical data derived from overburden in glaciated terrain is determining what part of the geochemical signal represents provenance, and what part represents cation redistributions resulting from post depositional geochemical processes. Of the post depositional (diagenetic) effects, weathering is by far the most significant. Since most geochemical exploration is carried out using samples from within or near the near-surface zone of weathering, low temperature geochemical processes can alter the provenance signal significantly. Low temperature geochemists – in effect most exploration geochemists- commonly try to take advantage of the redistribution and secondary concentration of cations by weathering; glacial sedimentologists generally try to avoid weathering effects as best as possible and use their geochemical results as surrogates for actual mineralogical composition that reflects provenance. Thus, it is critical to the latter group (and to the former as well), to understand to what extent vertical geochemical variations are provenance related and to what extent they are related to the local post depositional geochemical environment.

As in any detailed investigation, the conclusions drawn from this study can be grouped into three general classes; 1) site specific; 2) area-specific; and 3) independent or general (Shilts, 1978, p. 24-25).

1. *Site specific.* Site specific conclusions are those that are peculiar to the geological and physical setting of the site studied. They generally cannot be extrapolated away from the site itself.

A. No compositional or physical effects relating to late glacial northward flow were observed at this site. Compositional effects clearly related to flow reversal were documented at section 50, and possible effects of change in provenance were also observed at section 28.

B. Relative to other sections studied, the absolute concentrations of trace elements in these samples appear to be unique to this section, which is particularly enriched in Cu, Pb, Zn, Co, and Ni. Section 13, which is only a few kilometres from section 531, has metal concentrations far below those of 531.

2. *Area specific.* Area-specific conclusions are those that may apply to tills of the Quebec Appalachian region in general.

A. The high pyrite (sulphide) content of the sand and coarser sizes and the low carbonate content of till at section 531 are typical of tills throughout the Quebec Appalachians, at least between the United States border and the region southeast of Quebec City. The only exception is a region southwest of Sherbrooke where a large area of calcareous mudstone causes the overlying drift to be very carbonate-rich. The high concentration of pyrite reflects the almost universal dissemination of sulphides in the flyschoid bedrock of the region; the carbonate is largely debris from unmetamorphosed platform carbonates of the St. Lawrence Lowlands, transported southeastward across the Appalachians during the last glacial maximum. The high sulphide/carbonate concentrations give rise to other characteristics observable in till at section 531 and in typical Appalachian tills.

B. The marked colour contrast between tan-biege, oxidized till and gray to olive gray, unoxidized till is largely a function of staining by the abundant secondary iron oxides and hydroxides produced by oxidation of pyrite and other sulphides. The sharp colour boundary evident between oxidized and unoxidized till at section 531 is repeated everywhere in the region on well drained till surfaces and marks an abrupt chemical and mineralogical break. In this section and in others where carbonate concentrations are less than 10%, the depth of carbonate leaching is coincident with the depth of oxidation, although siderite grains were found in oxidized samples from section 531.

The 3-4 metre oxidation/leaching depth seems to be constant through this part of the Appalachians and is much greater than in other regions where sulphides can be found at depths just below the solum. A possible reason for the greater depths in the Appalachians might be that the sulphuric acid produced by the breakdown of pyrite enhances the further breakdown of pyrite throughout the zone of oxidation.

D. The general depletion of sulphides and chalcophile elements in heavy mineral separates from stream sediments is probably typical of this region. The unpredictable enrichment of some elements in the stream heavies is probably related to selective destruction of certain sulfide phases or to some secondary geochemical process that is poorly understood and not related in any obvious way to the mineralogy or chemistry of the till from which the stream sediment is largely derived (Shilts, 1976; Maurice, 1986).

D. The subhorizontal structures in till at section 531 are similar to those observed in till throughout the region. They probably indicate that the till is composed of debris bands that were melted out of the basal part of the Lennoxville glacier with little accompanying disturbance by meltwater.

E. The high concentrations of non-carbonate carbon in clay fractions of the till of section 531 are probably typical of till throughout the flyschoid belt of the Quebec Appalachians. The carbon is thought to derive from finely divided, amorphous carbon compounds in the local black slates and phyllites. The geochemical significance of the carbon is not known, but its role in nucleating gold precipitates from groundwater might be investigated.

3. *General.* General conclusions are those that are generally applicable to till composition studies, regardless of the geological or weathering environment.

A. Partitioning studies of sulphur and carbonate minerals show that the coarser fractions (mostly rock fragments >4 mm) are least affected by weathering processes. However, because of their relatively small number compared to numbers of grains in smaller size fractions, the composition of a single or a few clasts can have an inordinate effect on the apparent chemical or mineralogical composition, the so-called 'nugget effect'.

B. Partitioning studies indicate that carbonate minerals concentrate in the coarser fraction (>0.25 mm) relative to sand and silt sizes where they are diluted by large quantities of quartz and feldspar. In clay sizes they are diluted by phyllosilicate minerals. Within the fraction <0.25 mm, there are minor concentration peaks in coarse silt and fine sand. These results are in accordance with observations of Dreimanis and Vagners (1971). Sulphur is derived from abundant fine-sand to pebble size fragments of pyrite that were not easily reduced to smaller particles during glacial comminution because of the high hardness and noncleavable nature of the well-crystallized pyrite. Most other sulphides should behave similarly, but partitioning of easily cleavable phases such as galena or molybdenite should be studied further.

C. Standard geochemical analytical methods available from commercial laboratories can provide high levels of reproduceability judging from the close correspondance of analytical results of two sample suites collected across the base of the zone of oxidation in different years. Furthermore, sample processing and collection techniques used in this research, are adequate to reproduce vertical geochemical variations faithfully. The ambient physical and chemical conditions in the soil and oxidized C-horizon at this site did not lead to any appreciable drift in chemical characteristics of the normally reactive clay-size fraction of the till.

D. Total carbonate content data obtained using the Leco and Chittick methods are comparable, although the Leco method is faster and more economical to use. Total dissolution data are consistently higher for the same samples, and are considered in this study to be unreliable. Using the latter method, carbonate contents are determined on the basis of total weight loss after dissolution in HCl and washing of the residue to remove the CaCl precipitate. Because silt and finer sized minerals other than carbonate, such as Fe oxides and phyllosilicates, can be dissolved by HCl and removed by the washing processes, carbonate contents calculated can be artificially high for finest grain size fractions.

E. The weathering of Lennoxville Till at section 531 is characterized by a colour change from dark blue-grey or olive grey to yellow-brown, presumably an oxidation colour, accompanied by marked changes in mineralogy and chemistry. Concentrations of pyrite, trace elements in labile heavy minerals, carbonate, and sulphur all decrease markedly in the oxidized till, while concentrations of some trace elements in the clay size fraction increase. Similar patterns of trace element enrichment and depletion, particularly that of trace element decrease in oxidized till, are observed at other sections in the region.

These observations are particularly important for evaluating the usefulness of near-surface samples for geochemical exploration in glaciated areas. They indicate that any labile ore mineral incorporated into unoxidized drift is potentially liable to destruction above the water table. This means that direct sampling of oxidized drift (till or stratified ice-contact sediments) will yield heavy mineral samples depleted in labile ore minerals and in their geochemical signature. Even geochemical analyses of elements found in stable minerals, such as gold, tin, chromite, etc., may be difficult to interpret in oxidized drift samples if significant amounts of labile minerals are concentrated in unoxidized drift. This results from the over representation of stable minerals in samples from which labile minerals have been removed. Even geochemical analyses that employ the $<250\,\mu m$ or $<63\,\mu m$ fraction of drift can be affected by weathering, depending on the proportion of labile minerals, including carbonates, and the amount of sand and silt-sized material in the sample, for most of the labile minerals reside in these coarser grain sizes.

Sampling the coarse bulk or heavy mineral fraction of streams draining glaciated areas will produce results that may require careful evaluation to determine to what extent weathering has affected the drift banks from which they derive their sediment. Depletion of labile heavy minerals or enhancement of stable heavy minerals is particularly important in this environment because of the superimposed effects of fluvial concentration (by placering) of the heavier, often most resistant minerals. Secondary geochemical processes that concentrate or deplete trace elements in the fluvial environment are apparently also important, at least in the vicinity of section 531, in producing analytical results that differ from those that would be expected if only physical processes transfered minerals from drift to stream sediments.

In conclusion, the authors suggest that the results of this investigation, which have been compared with less detailed studies in other climatic and geologic environments in Canada (Keewatin, Shilts, 1975; Nova Scotia, Podolak and Shilts, 1977; New Brunswick, Kettles and Wyatt, 1984, for example), indicate that before sampling and analytical strategies are developed for mineral exploration in any glaciated area, a thorough understanding of the local relationship of vertical compositional variations to provenance and weathering is essential.

ACKNOWLEDGEMENTS

Many colleagues participated in this study at one time or another over the past decade, and the authors apologise in advance for possible omission of any of their names. We were assisted in profile sampling in the field by Mary Asselstine, R.N.W. DiLabio, W. Podolak, J. Adshead, Sharon Smith, Andrée Blais, and Stéphane Péloquin. Blais and Péloquin measured the till fabrics, and Blais did most of the heavy mineral counting. Adrienne LaRocque wrote preliminary technique descriptions and carried out some of the heavy mineral point count analyses and lithology counts. LaRocque also compiled much of the information and drew some diagrams. W. Podolak carried out much of the comparative carbonate and sulphur analyses assisted by Pauline Honarvar. All geochemical analyses were carried out by Bondar-Clegg Ltd. of Ottawa. The authors gratefully acknowledge the help of all these people.

REFERENCES

Chauvin, L., 1979. Dépôts meubles de la région de Thetford-Mines – Victoriaville. Ministère de l'Energie et des Ressources, Québec, DPV-622, 20 pp.
Clifton, H.E., R.E. Hunter, F.J. Swanson and R.L. Phillips, 1969. Sample size and meaningful gold analysis. United States Geological Survey, Professional Paper 625-C, 17 pp.
Dreimanis, A., 1962. Quantitative gasometric determination of calcite and dolomite by using Chittick apparatus. Journal of Sedimentary Petrology 32: 520-529.
Dreimanis, A. and U.J. Vagners, 1971. Bimodal distribution of rock and mineral fragments in basal till; In: R.P. Goldthwait (ed.), Till – A Symposium. Ohio State University Press: 237-250.
Elson, J.A., 1987. West-southwest glacial dispersal of pillow-lava boulders, Philipsburg-Sutton region, Eastern Townships, Quebec. Canadian Journal of Earth Sciences 24: 985-991.
Foscolos, A.E. and R.R. Barefoot, 1970. A rapid determination of total organic and inorganic carbon in shales and carbonates. Geological Survey of Canada, Paper 70-11, 14 pp.
Gadd, N.R., B.C. McDonald and W.W. Shilts, 1972. Deglaciation of Southern Quebec. Geological Survey of Canada, Paper 71-47, 19 pp.
Harron, G.A., 1976. Metallogénèse des gites de sulfures des Cantons de l'est. Ministere des Richesses Naturelles du Quebec, ES-27.
Kettles, I.M. and W.W. Shilts, 1983. Reconnaissance geochemical data for till and other surficial sediments, Frontenac Arch and surrounding areas. Ontario. Geological Survey of Canada, Open File 947.
Kettles, I.M. and P.H. Wyatt, 1985. Applications of till geochemistry in southwestern New Brunswick; acid rain sensitivity and mineral exploration. Geological Survey of Canada, Paper 85-1b: 413-422.
Lamarche, R.Y., 1971. Northward moving ice in the Thetford Mines area of Southern Quebec. American Journal of Science 271: 383-388.
Levinson, A., 1980. Introduction to exploration geochemistry. Applied Publishing, Wilmette, Illinois, 924 pp.

Lortie, G., 1976. Les écoulements glaciaires Wisconsiniens dans les Cantons de l'Est et la Beauce, Québec. Thèse de maitrise, McGill University, 200 pp.

Maurice, Y.T., 1986. Interpretation of a reconnaissance geochemical heavy mineral survey in the Eastern Townships of Quebec. Current Research, Part A, Geological Survey of Canada, Paper 86-1A: 307-317.

McDonald, B.C., 1967. Pleistocene events and chronology in the Appalachian region of southeastern Quebec, Canada. Unpublished Ph.D. dissertation, Yale University, New Haven, Connecticut, 161 pp.

McDonald, B.C. and W.W. Shilts, 1971. Quaternary stratigraphy and events in southeastern Quebec. Geological Society of America Bulletin 82: 683-698.

Merritt, R.S. and E.H. Muller, 1959. Depth of leaching in relation to carbonate content of till in central New York State. American Journal of Science 257: 465-480.

Paré, D.G., 1982. Applications of heavy mineral analysis to problems of till provenance along a transect from Longlac, Ontario to Somerset Island. Unpublished M.Sc. thesis; Department of Geography, Carleton University, Ottawa, Ontario.

Parent, M., 1987. Late Pleistocene stratigraphy and events in the Asbestos-Valcourt region, southeastern Quebec. Unpublished PhD thesis, Department of Geology, University of Western Ontario.

Podolak, W.E. and W.W. Shilts, 1978. Some physical and chemical properties of till derived from the Meguma Group, southeastern Nova Scotia. Geological Survey of Canada, Paper 78-1a: 459-464.

Rencz, A.N. and W.W. Shilts, 1980. Nickel in soils and vegetation of glaciated terrains. In J.O. Nriagu (ed.), Nickel in the Environment. John Wiley and Sons, Inc.: 151-188.

Shilts, W.W., 1973. Glacial dispersal of rocks, minerals, and trace elements in Wisconsinan till, southeastern Quebec, Canada. In R.F. Black, H.B. Willman and R.P. Goldthwait (eds.), The Wisconsinan Stage. Geological Society of America, Memoir 136: 189-220.

Shilts, W.W., 1975. Principles of geochemical exploration for sulphide deposits using shallow samples of glacial drift. Canadian and Metallurgical Bulletin, 68(757): 73-80.

Shilts, W.W., 1976. Glacial till and mineral exploration. In: R.F. Legett (ed.), Glacial Till. Royal Society of Canada, Special Publication 12: 205-233.

Shilts, W.W., 1978. Detailed sedimentological study of till sheets in a stratigraphic section, Samson River, Quebec. Geological Survey of Canada, Bulletin 285, 29 pp.

Shilts, W.W., 1981. Surficial geology of the Lac Mégantic area, Quebec. Geological Survey of Canada, Memoir 397, 102 pp.

Shilts, W.W., 1984. Till geochemistry in Finland and Canada. Journal of Geochemical Exploration 21: 95-117.

Slivitzky, A. and P. St. Julien, 1987. Compilation geologique de la region de L'Estrie-Beauce. Quebec Ministère de l'Energie et des Ressources, MM 85-04, 40 pp.

Thomas, R.H., 1977. Calving bay dynamics and ice sheet retreat up the St. Lawrence valley system. Géographie Physique et Quaternaire 31: 347-356.

# CHAPTER 12

# Field methods for glacial indicator tracing

HEIKKI HIRVAS & KEIJO NENONEN
*Geological Survey of Finland, Espoo*

## INTRODUCTION

With the needs of intensified ore prospecting, largely boulder tracing and geochemical mapping and investigations, in mind, a systematic till survey of Northern Finland was undertaken in 1972–1977 with the principal aim of determining the Quaternary stratigraphy, the directions of glacial flow and the transport distances of the till material. The area concerned is located at the centre of the region covered by the last glaciation in Scandinavia and has a surface area of approx. 78 000 km², thus accounting for about a quarter of the whole area of Finland. The survey was based on observations made in some 1400 test pits dug with a mechanical excavator, about 400 of which extended through the entire Quaternary mantle to reach the bedrock below. The pits were of a mean depth of 3.8 m, the deepest of all extending down for 8–11 m (Hirvas et al. 1977).

One result to emerge from the survey was that the till stratigraphy of the area was a good deal more complicated than had been expected and that there are places in Northern Finland in which evidence of at least five ice flow phases of different orientation, glacial stages and their corresponding till beds can be found. Examination of the transport distances of the till suggested that the majority of the material of stone and boulder size had travelled only a short distance to its present site, often only 1–3 km.

In 1978 this till stratigraphy survey was extended to Central and Southern Finland, where again 2 or 3 till beds of differing age and orientation were identified in many places (Hirvas and Nenonen 1987). The fact that glacigenic sediments representing different stages in the glaciation and different directions of ice flow at the time of their deposition have been preserved in the surficial deposits in Finland is of great significance as far as indicator tracing techniques are concerned. In the first place, the ore indicators contained in the till may have been transported glacially several times and also by the meltwater associated with the various glacial stages, giving rise to patterns of complex material transport (Hirvas et al. 1977). Secondly, the older till deposits and sorted sediments may well have covered the ore outcrops in places and protected them from later glacial erosion. This in turn will have prevented ore material from being incorporated into the younger till, that occurring nearer the surface, the ore traces in which will be largely redeposited and usually greatly diluted with the till covering the ore outcrop or with sorted sediments (Fig. 1).

The results obtained from the above projects demonstrated that efficient ore prospecting and geochemical research in glaciated terrains call for a thorough knowledge of the till stratigraphy and glacial geology of the area concerned. With this in mind, the

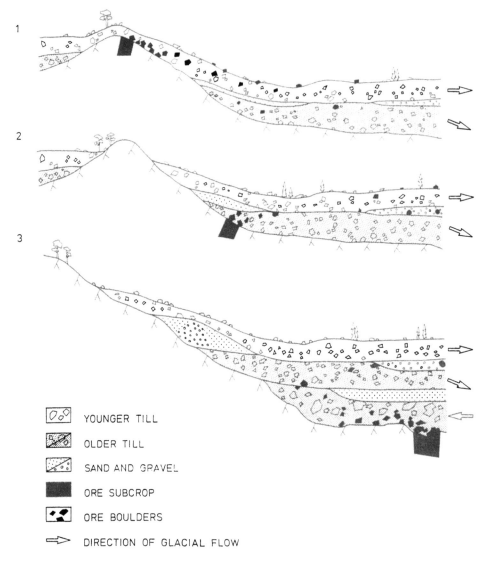

YOUNGER TILL

OLDER TILL

SAND AND GRAVEL

ORE SUBCROP

ORE BOULDERS

DIRECTION OF GLACIAL FLOW

Figure 1. Diagram illustrating the influence of glacial stratigraphy on the transport of ore indicators. Case 1: Indicators in the uppermost till have been detached by the last glacial flow. Case 2: Indicators found on the surface have been transported by two glacial flows and by meltwater – complex transport. Case 3: No ore indicators are to be found in the upper part of the sediment as the ore suboutcrop has been covered by several till or other deposits – glacially blind ore.

Quaternary Department of the Geological Survey of Finland set up a Quaternary Research Group in 1979, comprising three geologists and a research assistant, to serve the needs of ore prospecting. A further 4 or 5 geology students and a mechanical excavator are also engaged for the five-month fieldwork season, from May to September. The budget for this group, amounting to approx. US $ 250 000 (in 1988), represents about

1% of the money spent on ore prospecting annually in Finland. The purpose is to investigate the origins of reported ore indicators, erratics and geochemical anomalies by the methods of glacial geology in cooperation with the Exploration and Geochemistry Departments of the Geological Survey and other ore prospecting organizations. This group had investigated 57 ore indicators by December 1987 and had succeeded in locating the original bedrock deposit in 45 cases, which means 79%. Unfortunately none of these finds has yet led to commercial mining operations, but geological investigations and economic evaluations are still in progress at some of the sites. The aim of the present article is to describe the fieldwork procedures developed in connection with the above project, which is presented schematically in Figure 2. These methods have been adopted by mining companies in Finland and can be applied in all glaciated areas with thin or moderate till thicknesses, i.e. in the central parts of ancient continental ice sheets.

SELECTION OF THE AREA TO BE STUDIED

In order to ensure that the fieldwork proceeds smoothly and successfully, it is important to select the sites for investigation well in advance, so that sufficient time is available for planning the fieldwork, making the necessary preparations, choosing sites for bases and setting these up and other such preparatory stages. There are generally more sites or topics to be investigated than there is time for during the fieldwork season, and thus it is essential to select the sites in consultation with those who 'commission' the work, the geologists, geochemists and geophysicists working for the ore prospecting organizations. It is important to resist the temptation to select too many sites and to set aside sufficient time and other resources for a thorough investigation of each site chosen, so that the work can be carried through to a reasonable conclusion or intermediate stage. Such a result may not always be direct identification of the source of the indicator material, but may equally well be the assembly of sufficient geological observations and information to enable further investigations to be carried out or to justify abandonment of the topic. The actual choice of site for investigation, e.g. certain ore-bearing erratics, will be affected by numerous factors, such as:
  – The place where the erratics were found,
  – The surficial deposit and its genesis (basal till, esker, etc.),
  – The number of erratics, their size and roundness,
  – The depth at which they were found (on the surface or deeper down),
  – The thickness of the Quaternary cover (if known or deducible, e.g. from air photographs),
  – The type and potential size of the mineralization, i.e. the mineralogy of the erratics will indicate whether they are from only a small source rock, e.g. a single vein, or perhaps from a potentially economic mineralization,
  – The bedrock type and its potential in the immediate area,
  – The presence of cobbles or boulders which could be associated with the mineralized erratics,
  – Access routes (road network, mires, etc.),
  – Land ownership (in some cases),
  – An evaluation of the chances of success in tracing the source rock on the basis of the above information.

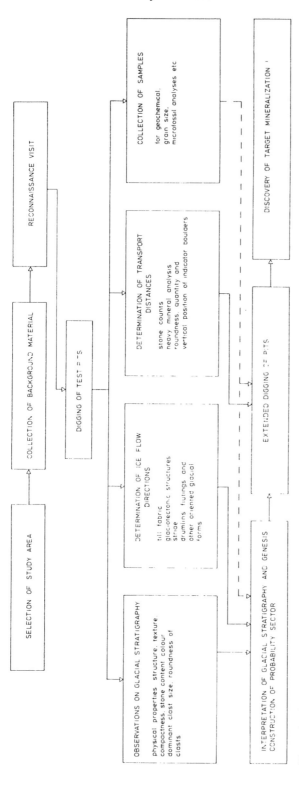

Figure 2. Diagrams representing the various stages in glacial indicator tracing.

## BACKGROUND INFORMATION AND RECONNAISSANCE

In order to evaluate the site and plan the research strategy, all the necessary information on the topic and area should be gathered from topographical, geological, geophysical and geochemical maps, stereo air photographs and geological papers on the area, for example. In addition to published data, reports stored in various archives and the geologists themselves who have studied the area previously often provide invaluable sources of information. Care in compiling preliminary information and preparing the work in advance can spare one many abortive steps in the course of the fieldwork. The material collected in this way will enable a preliminary interpretation of the geological conditions prevailing in the area to be formulated, including observations on ore-critical zones, the glacial geology, directions of glacial transport, transport conditions, etc. It is usually at this stage that the extent of the problem and the approach to be adopted becomes clear. The research group will be in close contact with exploration geologists, geochemists and geophysicists during this preparatory phase, the results of which should enable estimates to be made as to the time and resources required for the fieldwork.

A brief visit to the site should be made prior to the commencement of the fieldwork proper in order to formulate an accurate picture of the local geological conditions and the accessibility and topography of the area, etc. The routes to be followed by the excavator and the exact positions of the first pits to be dug should be marked out at this stage, and details of land ownership should be sorted out and the necessary permits obtained. Care taken to keep the owners of the land informed in an appropriately polite manner about what is going on can help to maintain the costs of compensation for the damage that will inevitably be caused at a moderate level. The majority of exploitable ore deposits in Finland at least have been discovered as a result of ore indicators found by the landowners themselves.

## DIGGING OF THE TEST PITS AND QUESTIONS OF SAFETY

The fieldwork will commence with the digging of a number of test pits at the site of the ore indicator, whether an erratic or a geochemical or heavy mineral anomaly. Since till is a non-homogeneous deposit and local conditions can vary greatly over a distance of twenty metres or so, the first pit to be dug should be located precisely at the indicator site.

The normal practice is to employ a mechanical excavator attached to a tractor (a backhoe) for the digging work (Fig. 3), on account of its mobility both in the field and on roads. An excavator of this kind will usually be able to dig down some 4–5 m, and can reach depths of up to 7 m by means of sloping one wall of the trench. The diameter of the top of the pit should be at least equal to the depth of the pit, in order to ensure a safe gradient of no more than 60° in the pit sides. Thicker Quaternary deposits and more extensive excavations require the use of a heavier excavator with a digging range of up to 8–9 m, or as much as 11 m with sloping technique. The moisture content, compaction and grain size of the sediment will affect the strength of the pit sides, and thus the safety of the digging operation, the most dangerous situations arising when digging in damp, loose, fine-grained material.

Working in deep exploration pits always entails a certain risk, which can be minimized by paying careful attention to safety factors such as the following (Fig. 4):

Figure 3. A tractor-mounted excavator offers a convenient means of constructing the explorations pits, especially in view of its mobility on ¬ads and in the field.

Figure 4. The exploration pits are usually damp, muddy places to work in. A helmet, reinforced boots, protective overalls and a ladder are all essential for safety purposes.

– Never work in a pit alone.

– Always leave one person on the edge of the pit to watch the sides while the others are working.

– Always wear a helmet and hard-toed protective boots, and protective goggles when cracking rocks.

– Ensure that those in the pit can leave it quickly. Use strong ladders.

– Check before going into the pit that there is no material in the sides that is likely to slip down, no loose stones or protruding material, and that any parts of the walls that could give way are smoothed away or cut in steps.

– Be careful of the moving excavator during the digging operation. Do not under any conditions use the scoop as a lift for getting in and out of the pit.

– Make sure that there is always a vehicle in attendance at the site and that first aid equipment is available.

– Do not enter a pit that looks in any way dangerous, as it will certainly prove to be dangerous.

– Make fence round the walls of deep, steep-sided pits.

Apart from actually digging the pits, the excavator can also be used for cutting various steps in the sides, clearing levels for stone counts and taking samples directly with its scoop if the pit proves too dangerous to enter.

OBSERVATIONS ON GLACIAL STRATIGRAPHY AND TILL PROPERTIES

The idea behind the tracing of ore indicators in surficial deposits is to follow the till bed containing the indicator back in the direction from which it was transported until the source rock is found. The following of such a bed naturally requires a knowledge of the manner of deposition of the till, its stratigraphy, its direction of transport and its transport distance.

The till stratigraphy and the genesis of the till unit containing the ore indicator should be determined using as wide a range of techniques as possible. In order to simplify and standardize such stratigraphical observations and render them better comparable one with another, a simple form was designed. The form contains a drawing of the profile (Fig. 5) and spaces for notes on the field analyses and samples taken from the pit and for the results of the analyses and physical properties of the stratigraphical units. Nowadays a more comprehensive computer form has been developed along the same lines. Use of a form of this kind will also ensure that all the necessary observations have been made.

Once the pit has been dug, the walls should be carefully cleaned with a spade and a bricklayer's trowel, as it is only from a properly cleaned pit wall that reliable observations can be made on the properties of the strata and the differences between them. Visual observations and determinations are made on the stratigraphical units to be detected and a profile is drawn of the most representative wall in the pit. Records are made of the physical properties of the till units, including colour, structure and grain size, and also the predominant (mean) degree of compactness, stone content, stone size and roundness of the stones ($\varnothing > 2$ cm) each on a scale of 1–5, where 1 denotes the smallest degree with respect to each property and 5 the largest (Kujansuu 1976) (Fig. 5).

If systematic differences in the above properties appear between several adjacent profiles and the strata can be compared from one pit to another, it is possible that the till beds concerned may be of different age or genesis. It should be remembered, however,

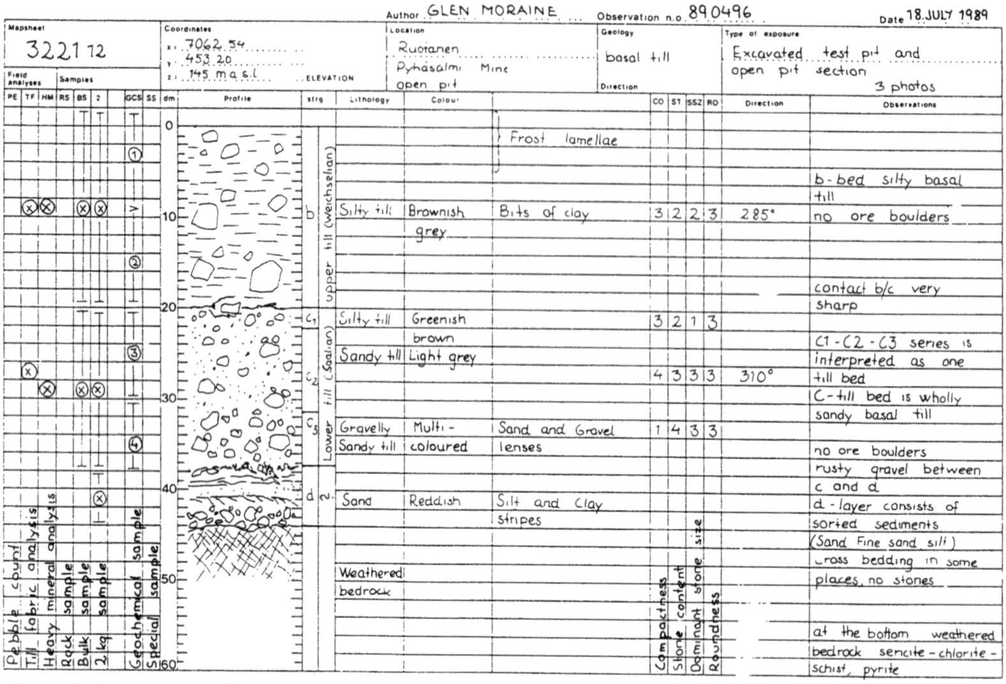

Author GLEN MORAINE    Observation n.o 89 0496    Date 18 JULY 1989

| Mapsheet | Coordinates | Location | Geology | Type of exposure |
|---|---|---|---|---|
| 3221 12 | x: 7062.54<br>y: 453.20<br>z: 145 m a.s.l. ELEVATION | Ruotanen<br>Pyhäsalmi Mine<br>open pit | basal till<br>Direction | Excavated test pit and<br>open pit section<br>3 photos |

Figure 5. Form for recording stratigraphical data from a test pit.

that a single till bed can vary in terms of certain properties on account of differences in the groundwater table (e.g. colour, compactness), proximity of the bedrock (e.g. stone size, stone content, roundness) or the thickness of the till bed (e.g. structure, compactness, stone content), for example.

Tills can be classified in terms of their physical properties as follows:

Compactness: 1. Extremely loose
2. Loose
3. Normal
4. Compact
5. Extremely compact, concrete-like

Basal tills will normally be of grades 2–4, (basal and ablation) melt-out tills of grades 1–2, and tills hardened by Fe, Mn, Ca or Si precipitations, and also certain 'old' weathered tills of grade 5.

Stone content: 1. Stoneless
2. Few stones
3. Normal
4. Abundant stones
5. Extremely abundant stones

Basal tills will normally be of grades 2–4, (basal and ablation) melt-out tills tend to vary greatly, tills composed of older fine sediments are of grades 1–2 and tills located close to the bedrock surface are sometimes of grades 4–5.

Stone size:    1. Small stones, < 6 cm
               2. Large stones, 6–20 cm
               3. Small boulders, 20–60 cm
               4. Boulders, 60–200 cm
               5. Large boulders, > 200 cm

Stone size is dependent on such factors as the physical properties of the source rock, transport distance, proximity of the bedrock surface and mode of transport, the most common grades in the shield areas of Finland being 2–3.

Roundness:    1. Non-eroded, sharp-edge, angular.
                 Clear fractured surfaces typical of individual rock types.
              2. Slightly eroded, slightly worn at edges, angular.
                 Still clear fractured surfaces typical of individual rock types.
              3. Eroded, edges eroded and rounded.
                 Original form still easily definable, fractured surfaces still retained.
              4. Rounded.
                 Original form difficult to define.
              5. Highly rounded.
                 Original form can no longer be defined.

Basal tills usually contain grades 2–4, melt-out tills vary greatly in this respect, tills composed of older glacio-fluvial or fluvial sediments contain grades 4–5 and short-distance transport basal tills grades 1–2, similarly frost-cracked surficial tills.

Careful definition of these properties will enable conclusions to be reached regarding the genesis or transport distance of the till in question. For example: Case 1, of a shear-structured till which was compact (4) and contained small boulders (3) which were slightly eroded (2), most probably represented a basal till, the stone material in which, including the ore indicators, had been transported over a short distance of not more than a few hundred metres. Case 2, a single, rounded, high-grade ore boulder, was found in a non-homogeneous, gravelly, normally compacted till (3) with large numbers (4) of small stones (1) which were highly rounded (5). Here the majority of the till was composed of redeposited older glaciofluvial material. The ore boulder had probably been transported some distance, a matter of kilometres, first by glacial melt-water streams and later by the glacial flow itself (complex transport).

Of crucial importance for determining till stratigraphy are analyses performed directly in the field, the results of which are available immediately, e.g. till fabric analyses, stone counts and heavy mineral determinations (for details, see below). These analyses should be performed systematically on each till unit identified by eye in the manner described above, and at the same time samples should be taken from each unit (in the form of groove samples extending through the whole depth of the stratum) for trace metal determination (at least 100 g) and grain size analysis (at least 250 g) in the laboratory. The purpose of the above field analyses is to enable a decision to be made as to which of the till beds identified by eye represent different deposits laid down during the same stage in glacial flow and which must be attributed to different stages in the glaciation. The results of the analyses to be carried out later in the laboratory will provide confirmation for such conclusions.

The heavy mineral analyses, stone counts, and till geochemistry, i.e. trace metal analyses, will point to the ore content of the till and thus provide important information for following the course of the till bed containing the ore. Ore-bearing boulders usually constitute the most concrete evidence of ore in a till, and these are often sufficient to allow an ore-bearing till bed to be traced to its source rock. Vigorous use of the geological hammer is a prerequisite for successful boulder tracing.

If interesting fine sediments or organic layers are to be found between the till beds, separate samples should be taken of these for microfossil analyses and radiocarbon dating. Such deposits, which may represent warmer interglacial or cooler interstadial periods between glaciations, occupy a key position when attempting to divide the lithostratigraphy, usually the till stratigraphy, into strata representing particular glaciations. If a key section for such organic deposits is to be found in or near the area being studied, attempts should be made to correlate the present till stratigraphy with the till bed contained in this.

DETERMINATION OF ICE FLOW DIRECTIONS; TILL FABRIC

In order to ensure success in tracing the ore indicator to its source, it is necessary to know precisely the directions of ice flow which may have influenced its course during transport. This assessment of ice flow directions should be carried out and confirmed on the basis of the widest possible range of evidence, from striae, drumlins, flutings and other oriented glacial landforms and glaciotectonic structures. Often the indicator train itself or other indicator trains in the area will serve to show the most recent direction of glacial flow. Since local directions of flow can deviate quite markedly from the general pattern, determinations should always be made on a local scale at and near the site of the indicator find. Deviant directions of ice flow are frequently found in zones of marginal and interlobate formations and in areas with a pronounced relief.

If no earlier results are available, the first information on flow directions will be obtained from the oriented glacial landforms visible on maps and air photographs of the area, these include radially oriented formations such as drumlins, flutings, crag and tail forms or eskers and transversely oriented ones such as various marginal and end moraines, push moraines and subglacially formed till ridges (c.f. Lundqvist, Kujansuu, this volume). This approach will usually suffice to provide a good general impression of flow directions in the area, which can then be filled out in the field. The weakness of this method is that there are some extensive areas which lack oriented glacial forms entirely. Another reason why one cannot rely entirely on maps and air photographs when determining ice flow directions is that in places such as the ice divide zone in northern Finland the erosional and depositional action of the most recent ice flow was so weak that it did not produce any oriented glacial landforms, and instead the drumlins laid down by an older flow 'show through' into the present-day surface topography. The discrepancy in flow direction between the two can be as much as 90°–130°. Sometimes the orientation and structure of the bedrock can prove deceptive in a similar manner when attempting to identify directions of ice flow from air photographs alone.

In areas with a thin or moderate till cover and an unbroken, glacially eroded bedrock surface, striae represent the most convenient and reliable means of determining directions of ice flow. The measured orientations of these striae again do not tell the 'whole truth' about the youngest direction of glacial movement, for example, as one may be dealing with ice diverted by an obstacle, an older direction of striation or even striae of some more recent origin, e.g. the action of modern sea or lake ice. Even the scoop of an excavator, snow chains of vehicles or stratification, schistosity, stretching or cracking in the bedrock can give rise to pseudo-striae, which one should be careful not to confuse with actual glacial striae.

Glaciotectonic structures can be used as a means of defining ice flow directions in some cases. The ice flow, ice push and shearing cause folds, overthrusts, faults, slip surfaces and capture structures, from which the direction of the tectonizing force, i.e. the direction of ice movement, can be measured.

The direction of ice flow can be ascertained directly from an ore-bearing till deposit most conveniently by means of till fabric analysis (first described by Holmes 1941), and this is without doubt the most important approach in areas that have no striated bedrock surfaces, oriented landforms or boulder trains. This method also provides an easy means of identifying till beds laid down by ice flows in different directions in the context of studying till stratigraphy.

The first essential requirement for a successful till fabric analysis is careful site selection. If the intention is to use the result of this analysis to determine the direction of glacial flow, it should be performed at a site which can be assumed to represent regional flow conditions. Relatively level areas of basal till are ideal for this purpose, and the flat distal summits of protrusions or gently sloping valleys are possible sites in more hilly terrain, but the slopes of hills, proximal sides of protrusions and steep-featured valleys and gorges should be avoided when considering regional glacial flow patterns as the plastic flow of the continental ice tends to follow even quite small features of the local topography at such points. It should also be remembered that solifluction can occur even at gradients of only 1°–2° (c.f. Boulton 1971, Downdeswell and Sharp 1986).

On the other hand, it is possible for a indicator boulder to have been transported only a few hundred metres from its bedrock source, perhaps less than a hundred metres, in which case purely local directions of ice flow or solifluction may have influenced its course. Thus it is always advisable to perform a till fabric analysis at the site where such an indicator was discovered or in the immediate vicinity.

Macrofabric analyses are performed on a horizontal surface, or 'fabric analysis table', cut into the wall of an existing section or a pit excavated in the till. This table should be of a size which allows the orientations of 100 elongated stones to be measured with ease, and should be located deeper than 1 m in order to avoid the effects of ground frost formation, solifluction, tree roots and other factors liable to interfere with the original orientation of the material. In cases of sandy or gravelly tills not susceptible to ground frost, fabric analyses can be performed at depths of only 0.5 m. Where thick till beds are concerned, fabric analyses should be made at a number of depths in order to reveal any changes in flow direction in the course of their deposition.

Each stone should be carefully removed from the till before measurement of its orientation in order to assess its true shape and the dip of its longitudinal axis. It should then be inserted back in its original position for measurement of the orientation or dip of its longitudinal axis to an accuracy of 5° with a geological compass. For a conventional

Figure 6. Each stone is removed from the fabric analysis table for determination of its longitudinal axis.

Figure 7. The stone is replaced for measurement of the direction of inclination of its longitudinal axis with a compass.

till fabric analysis all stones with a dip of over 30° may be denoted as vertical in attitude (Figs. 6 and 7).

The results should be drawn out in the form of a rose diagram in 10° orientation classes in terms of either absolute or percentage distributions (Fig. 8A), i.e. with the radial lines on the diagram drawn in proportion to the numbers of observations or percentage of all the observations falling within the given orientation class. Rose diagrams of this kind are simple to interpret on sight by imagining an ellipse formed by the peaks in the diagram, which will then represent the theoretical orientation of the basal till, on the assumption that the stones will have adopted their position in accordance with the flow mechanics of a viscous liquid (see Jeffrey 1922, Taylor 1923, Glen et al. 1957).

Elongated stones in basal tills are usually inclined in the direction of approach, due to the upward curvature of the surfaces over which they were sliding at the time of deposition or to other aspects of the flow mechanics (Fig. 8B). Saarnisto and Peltoniemi (1984) note on the basis of over a hundred till fabric analyses performed in the Kuhmo area of eastern Finland that about 60% of the stones in the till were inclined in the direction of approach of the glacial flow and 40% were inclined in the down-glacier direction or were not inclined at all.

Till fabric analysis is a statistical procedure, and consequently not every analysis will yield the true direction of glacial flow, or indeed any direction at all. This means that little weight can be attached to the results of a single analysis unless they are backed up by observations of striae, drumlins or flutings. Alternatively determinations based on till

**A**

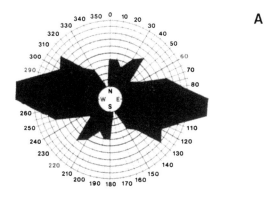

| | | 50 | 100 | | | 50 | 100 | % |
|---|---|---|---|---|---|---|---|---|
| 10 | |||| | 2 | 4 | 190 | | – | – | 4 | 4 |
| 20 | ||| | 2 | 2 | 200 | | | – | – | 3 | 3 |
| 30 | || | 1 | 2 | 210 | ||| | 2 | 3 | 5 | 5 |
| 40 | | | 1 | 1 | 220 | ||||| | 2 | 5 | 6 | 6 |
| 50 | | | 1 | 1 | 230 | | | – | 1 | 2 | 2 |
| 60 | | – | – | 240 | ||| | 3 | 3 | 3 | 3 |
| 70 | | | – | – | 250 | ||||| || | 4 | 7 | 7 | 7 |
| 80 | ||| | 1 | 2 | 260 | ||||| || | 2 | 7 | 9 | 9 |
| 90 | | | 1 | 1 | 270 | ||||| ||||| || | 5 | 12 | 13 | 13 |
| 100 | ||||| | 2 | 4 | 280 | ||||| |||| | 4 | 9 | 13 | 13 |
| 110 | || | 1 | 2 | 290 | ||||| || | 1 | 7 | 9 | 9 |
| 120 | | | 1 | 1 | 300 | ||||| | | 3 | 6 | 7 | 7 |
| 130 | |||| | 2 | 4 | 310 | ||||| | 3 | 4 | 8 | 8 |
| 140 | ||| | – | 2 | 320 | ||| | 1 | 2 | 4 | 4 |
| 150 | | | 1 | 1 | 330 | ||| | 1 | 2 | 3 | 3 |
| 160 | | – | – | 340 | | – | – | | |
| 170 | | – | – | 350 | | – | – | | |
| 180 | |||| | 2 | 4 | 360 | | – | – | 4 | 4 |

Total
Vertical

**B**

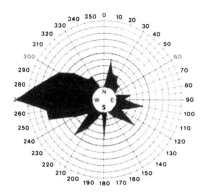

| | | 50 | 100 | | | 50 | 100 | ± |
|---|---|---|---|---|---|---|---|---|
| 10 | |||| | 2 | 4 | 190 | | | – | – | |
| 20 | || | 2 | 2 | 200 | | | – | | |
| 30 | || | | 2 | 210 | ||||| | 3 | 3 | |
| 40 | | | | | 220 | ||||| | 2 | 5 | |
| 50 | | | | | 230 | | | – | 1 | |
| 60 | | – | – | 240 | ||| | 3 | 3 | |
| 70 | | – | – | 250 | ||| || | 4 | 7 | |
| 80 | ||| | | 2 | 260 | ||| || | 2 | 7 | |
| 90 | | | | | 270 | ||||| ||||| || | 5 | 12 | |
| 100 | ||||| | 2 | 4 | 280 | ||||| ||| | 4 | 9 | |
| 110 | || | | 2 | 290 | ||||| || | 1 | 7 | |
| 120 | | | | | 300 | ||||| | | 3 | 6 | |
| 130 | |||| | 2 | 4 | 310 | ||||| | 3 | 4 | |
| 140 | ||| | – | 2 | 320 | | | 1 | 2 | |
| 150 | | | | | 330 | ||| | 1 | 2 | |
| 160 | | – | – | 340 | | – | – | |
| 170 | | – | – | 350 | | – | – | |
| 180 | |||| | 2 | 4 | 360 | | – | – | |

Total
Vertical

Figure 8. A. Till fabric assessment form and horizontal rose diagram. B. Till fabric stone inclination table and the corresponding rose diagram.

fabric evidence alone require a number of analyses to be made in the same area. Transverse orientation results can occur even at sites where no particular reason can be detected for this.

Other methods available for determining the orientation of till in addition to macro-fabric analysis are microscopic examination of the orientation of sand grains and the study of magnetic susceptibility anisotropy to determine the orientation of magnetic particles (cf. Seifert 1954, Harrison 1957, Evenson 1971, Boulton 1971, Gravenor and Stupavsky 1974, 1976, Puranen 1977). These methods naturally require the taking of undisturbed till samples in the field and their transfer to the laboratory for further treatment, e.g. for the preparation of polished sections. Since orientation data are required immediately in the field when engaged in ore prospecting, in order to plan the site for the next test pit, microfabric and magnetic fabric are, however, of little practical help.

When used correctly, till macrofabric analyses provide a convenient means for determining the direction of transport of an indicator boulder at the site of its discovery. It takes about 30 minutes for two people to produce a single till fabric analysis.

DETERMINATION OF TRANSPORT DISTANCES; STONE COUNTS

Once the position of the indicator boulder in the till stratigraphy and its direction of transport have been ascertained in the manner set out above, the remaining piece of information required to find the source rock is the transport distance. If this can be defined with accuracy, say to within a few hundred metres, one claim to have succeeded in taking a 'short cut' to the study of the mineralization.

One common means of defining the transport distance is through petrographic analysis of the till, i.e. stone counts. These entail the collecting of random samples of 100 or 200 stones and determining the percentages of the different rock types present. Definition of the petrographic composition of the till in this way and comparison of the results with the surrounding bedrock, usually as represented on a bedrock map, enables a maximum and minimum possible transport distance to be proposed for each rock type. Frequently a 'pathfinder' rock type presents itself which is associated with the bedrock formation containing the ore, so that common occurrence of this type in the stone counts will usually mean that the source rock is close at hand (a few hundred metres away or less). Parameters and measurement methods for the tracing of glacial transport distances are described in more detail by Bouchard and Salonen elsewhere in this volume.

The manner in which the stone counts are performed, in terms of the amount of sample used, fraction chosen and rock type classification to be employed, for instance, can be adjusted to the local conditions. It is common in Finland to concentrate on the fraction 6–20 cm, which is the mean stone size in tills and also frequently the size class into which the indicator boulders fall, but another common approach is to wash 10 litres of till sample through a sieve to obtain an unselected 2–6 cm diameter gravel fraction (Fig. 9) usually comprising several hundred stones, which are easy to identify to rock type in this washed condition and which can be readily carried back to the laboratory if necessary. In practice the rock type composition is virtually the same in all fractions, but the 6–20 cm fraction is to be recommended from the point of view of searching for indicator boulders (Fig. 10). Bouchard and Salonen (1988) have described a method for till lithological analysis using polished thin sections under a microscope. This involves a petrographic identification of 50 to 100 grains of coarse sand size (1–2 mm).

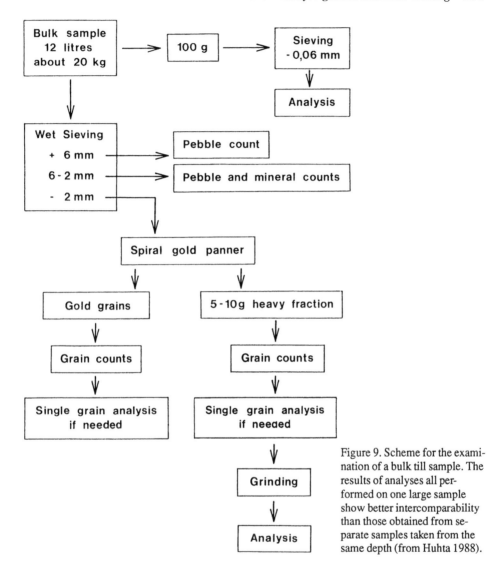

Figure 9. Scheme for the examination of a bulk till sample. The results of analyses all performed on one large sample show better intercomparability than those obtained from separate samples taken from the same depth (from Huhta 1988).

The material for a stone count, i.e. the selection of stones or till sample for sieving, can be taken from the till unit being studied, either directly from the various walls of the pit or from the material removed from the pit and piled up on the edge, although in the latter case it should always be ascertained that the stones really are from the till unit concerned and that digging has not advanced through the unit to older deposits below. It is important for the reliability of the result that the stones are collected systematically and without selection, care being taken to ensure that the indicator boulders themselves and their host rocks are not overrepresented and that the sample does not include indicator boulders recovered either earlier or later from the same pit. In general, care should be taken not to favour stones of a particular colour or shape or to concentrate consciously or unconsciously on the collection of rarities.

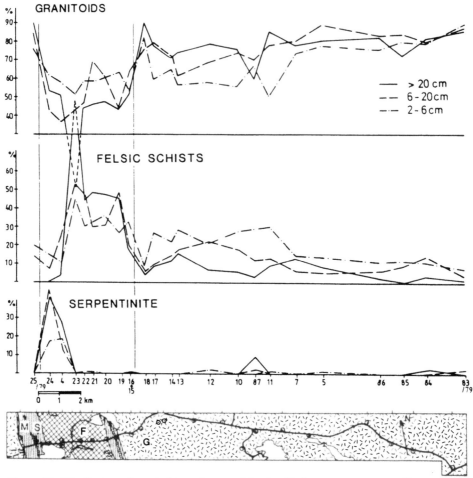

Figure 10. Distribution of granitoids and serpentinite in till along one survey profile across the greenstone belt in Kuhmo. Surficial boulders (>20 cm), stones (6-20 cm) and pebbles (2-6 cm) were examined separately. The ice flow was from left to right. Legend to the bedrock map: G = granitoids, F = felsic schists, M = mafic and ultramafic volcanite, S = serpentinite. The narrow serpentinite zone is immediately reflected in the till lithology. From Saarnisto and Taipale (1985).

Although the principles and methods outlined above enable the till stratigraphy and directions of glacial flow to be determined fairly accurately, there is no absolute, precise method in existence for determining the transport distance of a till or the ore indicators to be found in it. All the methods are to some extent or other estimations and their results either relative or average measures (Fig. 11). They are usually sufficient to allow one to say whether the transport distance is short or long (a few hundred metres or less vs. several kilometres) or to say whether the majority of the material has travelled more than or less than a certain distance. The problem arises from the fact that on account of their mode of formation, younger tills inevitably contain varying quantities of redeposited material from older tills and other sediments. Thus an individual ore boulder may have been transported to the site where it was discovered by a variety of processes of different age and operating in different directions (multi-stage transport), or even by processes differing in kind, e.g. glaciofluvial and glacial (complex transport).

A typical example of complex transport concerns the rapakivi erratics to be found in Southern Finland, which are derived from the cores of floating icebergs. These are thought to have been carried first in a NW–SE direction by the glacial flow and then in an E–W direction with the icebergs in the Baltic Ice Lake and finally, when these melted, to have been deposited in the till along with other surface boulders. Fortunately for those engaged in ore exploration, such cases of multi-stage and complex transport are fairly rare. The majority of the indicators that come to light are ones in which the material has travelled in only one direction, usually that of the youngest form of glacial transport, so that sufficiently reliable determinations or estimates can be made regarding the transport distance.

The landform containing the ore indicator can provide a clue to its transport distance, which will usually be long in the case of streamlined landforms, i.e. drumlins, fluting ridges and eskers, but shorter in that of transverse forms such as end moraines, push moraines and transverse basal moraine formations (Bouchard and Salonen in this volume). It should always be remembered when estimating the transport distance of ore indicators that the boulders concerned mostly represent only a small portion of the stone fraction of the till, while a tiny proportion of any till will have originated many hundreds of kilometres away, maybe over a thousand, e.g. the distribution of Precambrian material from Scandinavia in the tills covering the sedimentary rocks of northern Central Europe. An individual indicator boulder may therefore have travelled a considerable distance, or else it may be derived from a small ore outcrop which has produced very few boulders altogether for glacial transport. Multi-stage and complex transport are both common in cases of individual ore indicators.

In view of the difficulty of determining transport distances, it is worthwhile using a

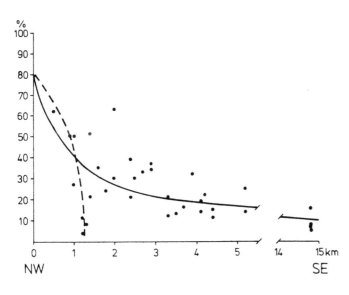

Figure 11. Variations in the percentage of gabbro pebbles in the till along the Akanvaara stone count transect, which runs parallel to the Early Weichselian glacial flow. Point 0 represents the contact between the gabbro and mica schists. The broken line shows the rapid decrease in gabbro content in the till lying closest to the bedrock face. From Hirvas et al. (1977).

Figure 12. Separation of heavy metals by panning. An experienced panner can process samples in this way faster and more accurately than a mechanical separator.

combination of methods for this purpose, including stone counts, frequencies of occurrence, numbers, sizes and roundness of ore boulders and their stratigraphical positions (i.e. their vertical distribution) in the till bed, heavy mineral determination and till geochemistry.

HEAVY MINERAL STUDIES

Recent results suggest that the determination of heavy mineral content is an extremely useful method for estimating the transport distances of ore boulders (see also Peuraniemi, this volume). In other words, when the heavy fraction of the till has proved to contain large amounts of the same ore minerals as the boulders whose origins are being determined, these boulders have usually proved to have been transported only a few hundred metres or less, whereas conversely, if the heavy till fraction has been devoid of the minerals contained in the ore boulders the latter have proved to be the result of long-distance transport, originating several kilometres away (c.f. Hirvas 1989).

The heavy mineral content of the till can also be used to guide further research. A small ore outcrop will usually give rise to only a few boulders, and these will be difficult to distinguish, especially when below the groundwater table. A till can often prove to be a flowing 'sludge' in which it is impossible to find all the ore boulders, whereupon ore mineral fragments from the outcrop can be reliably identified in the heavy mineral fraction, provided that they have not been entirely weathered away or dissolved (see Shilts and Kettles, this volume). This method is excellently suited for the tracing of oxidic ore minerals, metallic gold and sometimes certain sulphide minerals such as galena and chalcopyrite.

One advantage of heavy mineral determinations over trace metals, for example, is that it yields a result quickly, in the form of data on the numbers, sizes and roundness of the heavy mineral grains presents which can be used to reach conclusions on the distance from the source rock and direct the operations of the excavator from one day to the next.

Heavy mineral enrichments from tills should always be carried out using samples of a constant size, each of 12 litres or one bucketful of material, so that the results will be comparable (Fig. 9). The work can be done by panning by hand or mechanically, using a spiral gold panner or vibrating table, which yields high quality, readily comparable heavy mineral samples (Figs. 12 and 13).

Figure 13. A spiral enricher begins the separation process with the heavy mineral grains, yielding 'pure', easily examinable fractions. The reproducibility of the results obtained with this device is good. The smallest heavy mineral grains are largely washed out, e.g. the <0.02 mm gold fraction.

Heavy mineral determinations of this kind are of importance because there are extensive areas of northern Finland, for instance, where the bedrock surface, sometimes down to a depth of 100 m, consists of quite loose preglacial weathering products penetrable with a spade. Since the surfaces of any ore bodies in such areas will also be weathered in the same way, they will not produce boulders for incorporation in the local till but rather release finer-grained material, so that the ore indications have to be sought via heavy metal determinations or by geochemical methods.

SAMPLING BY DRILLING

Following the excavation stage, the search may be continued in various ways by drilling into the mineralized zone and the ore body itself. This approach offers the only possible means of locating the ore body in areas with a thick sediment cover, while boring into glacigenic formations enables samples to be obtained from precisely the desired depths.

The most widely employed and least expensive method is percussion drilling, samples being obtained by means of 'flow-through' bits. The drilling rods used are normally of length 100 cm and diameter 25 mm or 35 mm. This equipment enables samples of 150–300 g to be taken depending on the rod diameter and the length of the sampling bit.

The lightest drilling equipment available, weighing 10–25 kg, consists of a motor-driven percussion drill, e.g. Cobra, Pioneer or Partner, which can be carried into the field complete with its rods, but the trend nowadays seems to prefer a multi-purpose drill mounted on a light rig, in which case percussion drilling, raising of the rods and movement of the drilling beam all operate hydraulically. This is naturally more efficient and is capable of reaching greater depths, and allows excellent mobility in the field, the drill and rig weighing some 1000–2500 kg, but it requires motorized transport to the site and makes drilling in remote areas somewhat difficult. A modern multi-purpose drill of this kind can usually be used for percussion drilling, Auger boring and even deep drilling through Quaternary deposits and into the bedrock to depths of up to 60 m.

Percussion drilling from the bedrock surface can easily be used to produce a 'button sample' from which thin sections can be made.

The depth range of a portable percussion drill in sandy till with few boulders is 8–10 m and that of a rig-mounted multi-purpose drill 15–20 m using percussion drilling without flushing. Considerably greater depths can be reached by flushing with water and drilling through the stones in the till. Drilling equipment and samplers have developed so much in recent times that from a technical point of view no sites are impossible for sampling, the major limiting factor simply being cost. The presence of boulders and tightly compacted till horizons will slow down drilling but will not normally prevent it.

Drilling should take place at intervals of a few metres to a few hundred metres depending on the site, the points and transects being located so as to obtain the best possible section through the tectonic structures and stratification of the bedrock and mineralization and through any glacigenically transported indicator trains. Transects are best suited for studying already located mineralizations and networks of drilling points for mineralizations or anomalies of unknown shape and orientation in loose deposits.

Attempts should also be made to link the drilling samples with the stratigraphy of the Quaternary deposits, in which respect it is useful to take samples from a number of depths at at least some of the drilling points in order to ascertain the effect of the stratigraphy on variations in ore content between the horizons. A 'button sample' from the bedrock surface should always be accompanied by a sample from the overlying till, for the button sample represents the bedrock composition only at the sampling point itself, whereas the basal till 20 cm above it may well represent that of the formation in general, although in a somewhat weaker manner. Thus a bedrock or button sample alone taken by percussion drilling when investigating an ore type of a porphyritic, veined or breccia structure may provide an erroneous impression of the mineralization.

It is customary to repeat percussion drilling several times with small adjustments in the position of the equipment in order to avoid till boulders that may obstruct the passage of the bit and obtain a sample from the desired depth. Finally, it should be remembered that the quality of the samples is more important than the number of holes drilled in a day. Coker and DiLabio (1988) have summarized the properties of drilling devices currently in use in drift prospecting in Canada (Table 1).

PROCESSING OF SAMPLES

The 100–200 g till samples from the pits and drillholes intended for trace metal analysis should be air dried in an oven and divided by sieving into a fine i.e. silt + clay fraction of < 0.06 mm, a medium fraction of 0.06–0.5 mm and a coarse fraction of > 2.0 mm. Usually it is only the fine fraction that is used for trace metal analysis. Sometimes also the clay fraction (less than 0.002 mm) is separated from the till matrix and used for chemical analyses (see Shilts and Kettles, this volume). Procedures for major and trace element analyses are routine in geochemical laboratories and they are outside the scope of the present paper.

Heavy mineral enriched fractions from bulk samples of 20 kg can be analysed for individual elements by the same methods using powdered material (see Fig. 9). Other available analytical methods can be used as the need arises.

The grain size distribution of the till should be analysed from samples of at least 200 g by sieving the material of grain size over 0.06 mm after washing and drying. The finer fraction can be analysed areometrically. It is common nowadays to use an automatic particle analyser based on accelerated sedimentation and the use of X-ray radiation. Computer methods can be used to draw out a grain size distribution curve, to calculate the grain size classes as percentages by weight and to obtain a wide range of grain size indices, e.g. mean grain size, sorting (McCammon 1962), skewness and kurtosis (Folk and Ward 1957). Corresponding grain sizes should be indicated for the sieving percentages Dm 25%, Dm 50% and Dm 75% (see Fig. 14). Usually only the median grain size, Dm 50%, and the sorting value are relevant to practical indicator tracing in till investigations, since it is these that tell us adequate information about the till deposition conditions. A surficial ablation till, for example, will usually have a coarse grain size distribution and be better sorted that the basal tills in the same area (cf. also Dreimanis, this volume).

Table 1. Features of various overburden drilling systems (averages based on 1985 data, costs in $ CAN) (from Coker and Dilabio 1989).

| | Reverse circulation drills (Longyear or Acker) (Nodwell mounted) | Rotasonic drills (Nodwell or truck mounted) | Small percussion and vibrasonic drills (various) | Auger drills (various) |
|---|---|---|---|---|
| 1. Production cost estimate per: | | | | |
| – day (10 hrs) | $1,800-$2,000 | $3,000-$4,000 | $500-$1,000 | $800-$1,500 |
| – metre | $25-$40 | $50-$80 | $20-$40 | $20-$40 |
| 2. Penetration depth | Unlimited (125 m?) | Unlimited (125 m?) | 10-20 metres (greater?) | 15 to 30 metres (boulder free) |
| 3. Environmental damage | 5 metre wide trails (may have to be cut in areas of larger trees) | 5 metres wide cut trails | nil | 2-3 metre wide cut trails (Nodwell, muskeg, all terrain vehicle mounted quite manoeuverable) |
| 4. Size of sample | 5 kg (wet) | Continuous core | 300 g (dry), or continuous core | 3-6 kg (dry or wet) |
| 5. Sample of bedrock | Yes (chips) | Yes (core) | Yes (chips) if reached | Unlikely, if hollow auger, split spoon sampler can be used for chips |
| 6. Sample recovery | | | | |
| a) till | Good | Excellent | Good | Good |
| b) stratified drift | Moderate | Excellent | Good | Poor to moderate |
| 7. Holes per day (10 hrs) | 4 @ 15-20 metres / 1 @ 60-80 metres | 4 @ 15-20 metres / 1 @ 60-80 metres | 5 @ 6 to 10 metres | 1 to 3 @ 15 to 20 metres |
| 8. Metres per day (10 hrs) | 60-80 metres | 60-80 metres | 30 to 50 metres | 20 to 60 metres |
| 9. Time to pull rods | 10 min @ 15 metres | 10 min @ 15 metres | 30 to 60 min @ 15 metres | 20 to 40 minutes @ 15 metres |
| 10. Time to move | 10-20 minutes | 15-30 minutes | 30 minutes | 15 to 60 minutes |
| 11. Negotiability | Good | Moderate | Good (poor if manually carried on wet terrain) | Good to reasonable |
| 12. Trails required | Yes, may have to be cut in areas of larger forest | Yes, must be cut | No | Yes and no |
| 13. Ease in collecting sample | Good | Excellent, continuous core | Sometimes difficult to extract from sampler | Good (contamination?) |
| 14. Type of bit | Milltooth or tungsten carbide tri-cone | Tungsten carbide ring bits | Flow through sampler, continuous coring | Auger with tungsten carbide teeth |
| 15. Type of power | Hydraulic-rotary | Hydraulic-rotasonic | Hydraulic percussion (gas engine percussion, vibrasonic) | Hydraulic-rotary |

Table 1. Features of various overburden drilling systems (averages based on 1985 data, costs in $ CAN) (from Coker and Dilabio 1989) (cont.)

| | Reverse circulation drills (Longyear or Acker) (Nodwell mounted) | Rotasonic drills (Nodwell or truck mounted) | Small percussion and vibrasonic drills (various) | Auger drills (various) |
|---|---|---|---|---|
| 16. Method of pulling rods | Hydraulic | Hydraulic | Hydraulic jack, hand jack or winch | Winch or hydraulics |
| 17. Ability to penetrate boulders | Excellent. Reverse circulation drills (Longyear or Acker) (Nodwell mounted) | Excellent, cores bedrock. Rotasonic drills (Nodwell or truck mounted) | Poor. Small percussion and vibrasonic drills (various) | Poor to moderate. Auger drills (various) |
| 18. Texture of sample | Slurry (disturbed sample) | Original texture (core can be shortened, lengthened and/or contorted) | Original texture | Original texture (dry) to slurry (wet) |
| 19. Contamination of sample | Nil, fines lost (tungsten) | Nil (tungsten) | Nil (tungsten) | Nil to high (tungsten) |

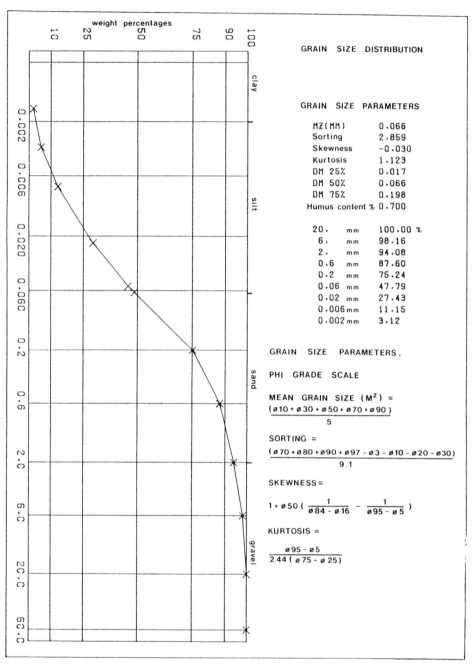

Figure 14. A typical granulometric curve for silty till in Finland and the parameters to be defined from it. Note: Millimetre scale sieves are nowadays most often used. The equivalent Tyler standard mesh-values for some commonly used sieves are as follows: 0.038 mm – 400 mesh, 0.044 mm – 325 mesh, 0.063 mm – 250 mesh, 0.125 mm – 115 mesh, 0.177 mm – 80 mesh, 0.250 mm – 60 mesh, 1.0 mm – 16 mesh, 2.0 mm – 9 mesh.

INTERPRETATION OF THE FIELD DATA

*Glacial stratigraphy*

Once all the necessary field observations have been made one can begin the task of interpreting the till stratigraphy. The most essential point for tracing the ore indicator is to be able to recognize till beds deposited under flow regimes operating in different directions and to distinguish the various genetically different till and other horizons. Glaciated areas will often feature at least say two till beds of different ages, implying that the ore indicator being investigated may have been transported during at least two stages and frequently also in at least two directions. This means that the site has to be examined sufficiently carefully that one can be quite sure which stratigraphic unit the indicator belongs to and from which direction it has come. If one cannot be certain about these points it will be impossible to follow the transport path of the indicator boulder back to the source rock.

Assistance in interpreting the stratigraphy can often be obtained from a resolved key stratigraphy for a nearby site, with which the till and other horizons identified at the site under investigation can be compared (Fig. 15). Attempts should also be made to trace till horizons with similar characteristics from one section to the next, as this may enable a feature detectable in one profile, e.g. a sorted sediment or erosion contact typical of an ice-free period, to be used to solve a stratigraphical problem (Fig. 16). The aim is then to set out to trace the transport path of the till horizon containing the ore indicator (boulders or heavy minerals) in an upglacier direction by means of excavations or percussion drill sampling.

*Construction of probability sector*

A convenient means of delimiting the most promising area for further investigation is to construct a probability sector for the glacial transport of the proximal boulder in the indicator train or whatever other ore indicator has been found in the till (Hirvas 1981, Nenonen 1984). The apex of the sector will consist of the proximal tip of the indicator train or whatever other glacigenically transported indicator one is studying, and its lateral bounds will be determined by the direction of the glacial flow responsible for the transport or the limits of variation in the direction of this flow. One should remember, of course, to allow for the declination between true and magnetic north when transferring the orientation measurements to the map or using map directions in the field.

In the indicator tracing example shown here the fan of zinc ore boulders, the local drumlin and striae orientations and the till fabric all lie within the arc 280°–300° * i.e. from W–NW and the edges of the probability sector are determined accordingly (Fig. 17). These lines can in theory be continued back as far as one likes, but in practice there is usually some geological, geophysical or geochemical boundary from beyond which the indicator could not have originated. Thus the probability sector serves in practice to delimit the area within which the indicator must have broken away from its source rock,

---

*It is a common practice to indicate the glacier flow direction as a direction of the glacier approach, like wind direction, but also the direction towards which the glacier flows is sometimes given.

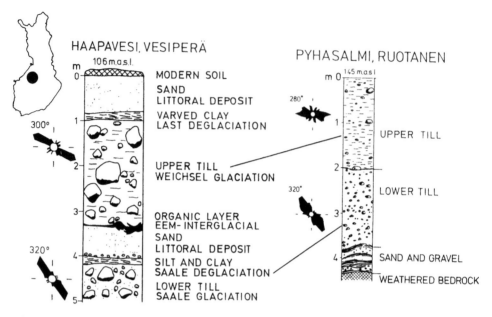

Figure 15. An example of the use of a key stratigraphy for correlating regional lithostratigraphical units.

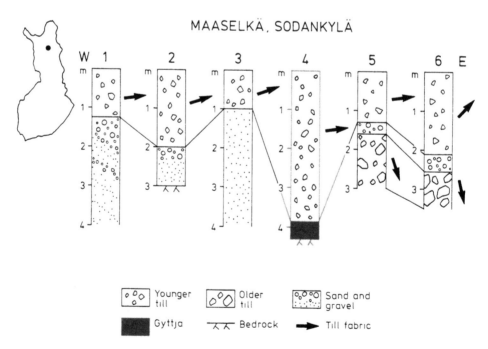

Figure 16. An example of the matching of till stratigraphies from one section to another by reference to the till fabric, sorted and organic interlayers and physical properties of the till.

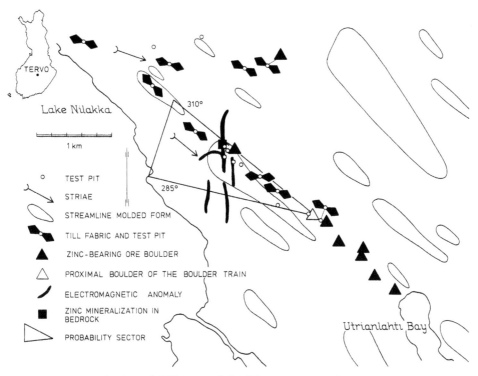

Figure 17. An example of a probability sector defined for target prospecting purposes.

assuming that it was transported to the site where it was discovered by the glacial flow depicted in the sector. If, as the research progresses, a new, more proximal glacigenically transported ore indicator of the same kind is discovered within the sector, a new sector can be constructed with this as its apex, so that the area to be studied is restricted further (Fig. 18). The normal pattern is thus that additional research gradually leads to a closer specification of the direction of glacial transport and a narrowing of the probability sector. At the same time geological, geophysical and geochemical maps of the area within the sector are usually available which enable part of it to be ruled out or even point directly to the most likely location of the source rock (Fig. 18).

A probability sector of this kind can also enable parts of known ore-critical geophysical and geochemical horizons or geological sequences to be conveniently delimited for further investigation, or alternatively it may serve to reveal the ore-critical nature of a certain zone.

The probability sector technique works best in cases of relatively short-distance transport, less than 10 km, where the area left within the sector is sufficiently small to allow efficient planning of the next stage in the investigation (Fig. 19). Where the indicator has been transported by a number of glacial flows, however, the sector tends to be broad and the area to be studied vast. Then once one succeeds in finding ore indicators attributable to the older till, one is on the track of the primary indicator train and the area to be investigated can be cut down considerably.

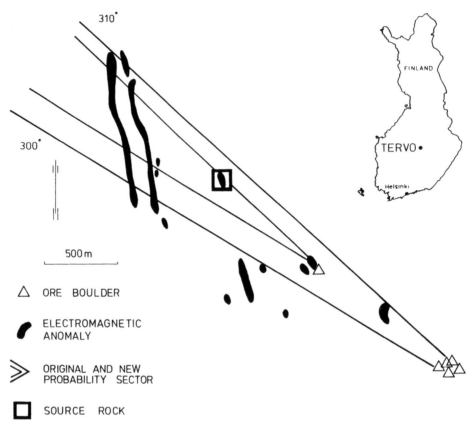

Figure 18. As the investigation proceeds, new sectors are defined with new indicator finds at their tips, thus progressively reducing the target area.

If the source rock cannot be found in the sector defined as above, it may be assumed that either 1) it was transported to the site by a flow in a direction not known when constructing the sector, 2) it was not transported there glacigenically, or 3) the source rock is particularly difficult to locate on account of long-distance transport or a highly localized occurrence of the rock itself, or for some other reason.

### Demarcation of the source rock

Having constructed a probability sector, the research team will proceed to investigate the area lying within it by digging further test pits on the side of the proximal boulder corresponding to the probable direction of glacial approach. If the boulder fan or probability sector is narrow and the boulders are thought to be of local origin, having been transported only a few hundred metres of less, one may prefer to dig inspection trenches across the sector at an angle that cuts the assumed direction of glacial flow. This will reveal the course of even the narrowest of boulder fans, which could be missed by digging pits if the network is not sufficiently dense. Any new ore-bearing boulders or

# VIITASAARI, Niinilahti

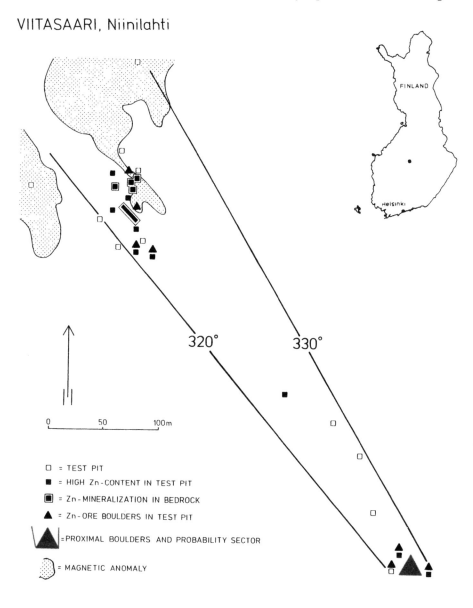

Figure 19. An example of the use of the probability sector strategy to locate the source rock of ore indicators which have travelled only a few hundred metres on the basis of a geophysical anomaly.

heavy mineral traces in the till will enable the digging to be directed more precisely towards the source rock. The frequency of occurrence of such evidence can be expected to increase as one approaches the outcrop, the boulders will be larger and more angular and they will be found progressively deeper down in the till bed concerned. At the same time the indicator train will often disappear entirely from the surface of the ground, 'dipping down' towards its source rock. The heavy metal content of the till will naturally increase towards the source, and the heavy fraction will usually contain increasingly large, angular grains and higher proportions of multi-mineral grains representative of the composition of the mineralization.

As soon as no more ore boulders or heavy metals representing the mineralization are to be found it is obvious that digging has advanced beyond the source rock, and it is necessary either to turn back and continue the search until the outcrop is found or to return to the point at which the transport path disappeared.

The excavation work at the final stages is also governed to a great extent by the stratigraphic and tectonic structures of the bedrock, as one attempts to obtain as good a cross-section of the bedrock and the mineralization as possible.

In areas with a thick cover of surficial deposits an attempt has to be made first to pin the outcrop down to as small an area as possible, where it can be located more precisely by e.g. percussion drilling. Characterization of the mineralization itself then normally proceeds by means of deep drilling and more precise petrological, geophysical and geochemical investigations. The purpose of studying the glacial geology of Quaternary deposits is to assist in directing these more expensive and more time-consuming investigations towards the correct target quickly, cheaply and with maximum certainty.

REFERENCES

Bouchard, M.A. and V.-P. Salonen, 1988. Till composition from granule thin-section analysis. Geologi 40: 159–161.

Boulton, G.S., 1971. Till genesis and fabric in Svalbard, Spitsbergen. In R.P. Goldthwait (ed.), Till – A symposium. Ohio State University Press, Columbus: 41–72.

Dowdeswell, J.A. and M.J. Sharp, 1986. Characterization of pebble fabrics in modern terrestrial glacigenic sediments. Sedimentology 33: 699–710.

Coker, W.B. and R.N.W. DiLabio, 1989. Geochemical exploration in glaciated terrain: geochemical responses. Exploration '87. Ontario Geological Survey, Special Volume 3: 336–383.

Evenson, E.B., 1971. The relationship of macro- and microfabric of till and genesis of glacial landforms in Jefferson County, Wisconsin. In R.P. Goldthwait (ed.), Till – A symposium. Ohio State University Press Columbus 345–364.

Glen, J.W., J.J. Donner and R.G. West, 1957. On the mechanism by which stones in till become oriented. American Journal of Science 255: 194–205.

Folk, R.L. and W.C. Ward, 1957. Brazos River Bar: a study in the significance of grain size parameters. Journal of Sedimentary Petrology 27: 3–26.

Gravenor, C.P. and M. Stupavsky, 1974. Magnetic susceptibility of the surface tills of southern Ontario. Canadian Journal of Earth Sciences 11: 658–663.

Gravenor, C.P. and M. Stupavsky, 1976. Magnetic, physical and lithologic properties and age of tills exposed along the east coast of Lake Huron, Ontario. Canadian Journal of Earth Sciences 13: 1655–1666.

Harrison, P.W., 1957. A clay-till fabric: its character and origin. Journal of Geology 65: 275–308.

Hirvas, H., 1981. Application of till stratigraphical studies in ore prospecting. In G.I. Gorbunov, B.I.

Koshechkin and A.I. Lisitsyn (eds.), Glacial deposits and glacial history in eastern Fennoscandia. Academy of Sciences of the USSR. Apatity: 48–54.

Hirvas, H., 1989. Application of glacial geological studies in prospecting in Finland. Geological Survey of Canada, Paper 89–20: 1–6.

Hirvas, H., A. Alfthan, E. Pulkkinen, R. Puranen and R. Tynni, 1977. Raportti malminetsintää palvelevasta maaperätutkimuksesta Pohjois-Suomessa vuosina 1972–1976. Summary: A report on glacial drift investigations for ore prospecting purposes in northern Finland 1972–1976. Geological Survey of Finland, Report of Investigation 19, 54 pp.

Hirvas, H. and K. Nenonen, 1987. The till stratigraphy of Finland. In R. Kujansuu and M. Saarnisto (eds.), INQUA Till Symposium, Finland 1985. Geological Survey of Finland, Special Paper 3: 49–63.

Holmes, C.D., 1941. Till fabric. Geological Society of America Bulletin 49: 1299–1354.

Huhta, P., 1988. Studies of till and heavy minerals for gold prospecting at Ilomantsi, eastern Finland. In D.R. MacDonald (ed.), Prospecting in areas of glaciated terrain. The Canadian Institute of Mining and Metallurgy: 285–292.

Jeffery, G.P., 1922. The motion of ellipsoidal particles immersed in a viscous fluid. Royal Society of London Proceedings A 102: 161–179.

Kujansuu, R., 1976. Glaciogeological surveys for ore prospecting purposes in Northern Finland. In R.F. Legget (ed.), Glacial Till. The Royal Society of Canada, Special Publication 12, Ottawa: 225–239.

McCammon, R.P., 1962. Efficiences of percentile measurements for describing the mean size and sorting of sedimentary particles. Journal of Geology 70: 453–465.

Nenonen, K., 1984. Till stratigraphic studies as an aid to ore prospecting in Finland. Striae 20: 101–105.

Parasnis, D.S., 1973. Mining geophysics. Elsevier, Amsterdam, 395 pp.

Puranen, R., 1977. Magnetic susceptibility and its anisotropy in the study of glacial transport in northern Finland, Prospecting in areas of glaciated terrain. Institution of Mining and Metallurgy, London: 111–119.

Saarnisto, M. and H. Peltoniemi, 1984. Glacial stratigraphy and compositional properties of till in Kainuu, eastern Finland. Fennia 162: 163–199.

Saarnisto, M. and K. Taipale, 1985. Lithology and trace-metal content in till in the Kuhmo granite-greenstone terrain, Eastern Finland. Journal of Geochemical Exploration 24: 317–336.

Seifert, G., 1954. Das mikroskopische Korngefüge des Geschiebemergels als Abbild der Eisbewegung, zugleigh Geschichte des Eisabbaues in Fehrmarn, Ost-Wagrien und dem Dänischen Wohld. Meyniana: Veröffentlichungen aus dem Geologischen Institut der Universität Kiel. Band 2: 126–184.

Taylor, G.I., 1923. The motion of ellipsoidal particles immersed in a viscous fluid. Royal Society of London Proceedings A 103: 58–61.

APPENDIX: GEOPHYSICAL METHODS

Risto Puranen
*Geological Survey of Finland, Espoo*

Geophysical methods have traditionally been used to locate and delineate objects hidden within and below the overburden. In the case of glacial indicator tracing these objects are boulders or anomalous layers of till and their sources in the bedrock. In order to be geophysically identified, the objects must differ from their surroundings in terms of density or magnetic, electric, elastic or radiometric properties. The surroundings of the objects can also be outlined by profiling the thickness of the overburden and the topography of the bedrock by geophysical methods.

Cross-sections of the overburden have generally been studied by means of seismic and resistivity profiling. With normal efforts these methods lead to rather crude profiles, which show large gaps between data points, but the seismic method in particular can penetrate a very thick overburden almost independent of the soil type involved. A more recent profiling apparatus is the electromagnetic impulse or ground-probing radar, which is well suited for continuous, rapid investigations of till cross-sections under Scandinavian conditions (Ulriksen, 1982). Impulse radar can be used to map the topography of the bedrock accurately under a few meters of till and it is sometimes also possible to locate boulders within the till cover. The weak point of this radar technique is its disability to penetrate more than a few meters of electrically conductive clay.

When searching for sulphide ore boulders it is common to use various metal detectors and light-weight (electro)magnetometers. These can be used to discriminate between highly conductive and magnetic boulders under favourable conditions. The radioactive boulders are traced with the aid of scintillometers and gamma spectrometers. The use of spectrometers is preferable as they characterize the whole energy spectrum of gamma radiation, which defines the source more specifically. Electromagnetic and radioactivity meters enable anomalous boulders to be located only when they are some tens of centimeters below the till surface. Several versions of these metres are commercially available, and their use and applications are presented in handbooks published by the manufacturers.

The train of anomalous boulders leads towards the source formation, but often comes to an end before the source is reached. Anomalous elements therefore need to be located in finer till fractions within the overburden. The radioactivity of finer fractions can be measured directly in investigation pits (Österlund, 1982) and it is also possible to measure the magnetic susceptibility, reflecting the magnetite content of the till, from the pit walls (Puranen, 1977). The amounts of radioactive and magnetic elements can thus be traced in the field during sampling for later laboratory measurements.

The anisotropy of magnetic susceptibility in oriented till samples can be measured in the laboratory. This characterizes the orientation of magnetic minerals in the till (Stupavsky et al., 1974), which then gives an estimate for the direction of glacial transport. The bulk susceptibility of the till reflects the amount of glacigenically transported magnetite (Pulkkinen et al., 1980), and the grain density of the till samples gives a preliminary estimate for the main mineral composition of the till (Puranen and Kivekäs, 1979). Finally, it is especially useful to determine a petrophysical 'fingerprint' for the boulder samples by means of laboratory measurements, as this will facilitate the identification and delineation of the source formations on geophysical maps (Ketola et al., 1976).

## References

Ketola, M., M. Liimatainen and T. Ahokas, 1976. Application of petrophysics to sulfide ore prospecting in Finland. Pure and applied geophysics 114: 215–234.

Österlund, S.-E., 1982. Micro boulder tracing with a portable gammaspectrometer. Prospecting in areas of Glaciated Terrain. The Canadian Institute of Mining and Metallurgy 240–248.

Puranen, R., 1977. Magnetic susceptibility and its anisotropy in the study of glacial transport in northern Finland. Prospecting in areas of glaciated terrain. Institution of Mining and Metallurgy, London: 111–119.

Pulkkinen, E., R. Puranen and P. Lehmuspelto, 1980. Interpretation of geochemical anomalies in glacial drift of Finnish Lapland with the aid of magnetic susceptibility data. Geological Survey of Finland, Report of Investigation 47, 39 pp.

Puranen, R. and L. Kivekäs, 1979. Grain density of till: a potential parameter for the study of glacial transport. Geological Survey of Finland, Report of Investigation 40, 32 pp.

Stupavsky, M., C.P. Gravenor and D.T.A. Symons, 1974. Paleomagnetism and magnetic fabric of the Leaside and Sunnybrook tills near Toronto, Ontario. Geological Society of America Bulletin 85: 1233–1236.

Ulriksen, P., 1982. Application of impulse radar to civil engineering. Lund University, Department of Engineering Geology. Doctoral Dissertation. 179 pp.

# Index

249

Milton Keynes UK
Ingram Content Group UK Ltd.
UKHW020822141024
449569UK00008B/518